Science Cycles

Grades 3-6

By
Sharon Sivinski

Cover Illustration by
Matthew Van Zomeren

Published by Instructional Fair • TS Denison
an imprint of

McGraw Hill Children's Publishing

Author: Sharon Sivinski
Editors: Debra Olson Pressnall, Angella Phebus, Mary Rose Hassinger
Cover Illustration: Matthew Van Zomeren
Inside Illustrations: Kathryn Marlin
Cover Photos: ®Corel Corporation

McGraw Hill Children's Publishing

Published by Instructional Fair • TS Denison
An imprint of McGraw-Hill Children's Publishing

Send all inquiries to:
McGraw-Hill Children's Publishing
3195 Wilson Drive NW
Grand Rapids, Michigan 49544

Science Cycles—grades 3–6
ISBN: 0-7424-0248-7

2 3 4 5 6 7 8 9 10 11 12 PHXBK 08 07 06 05 04 03

The McGraw-Hill Companies

Table of Contents

Introduction

While teaching science at the New Mexico Museum of Natural History, I became aware that children need stories to understand big ideas. The stories they hear can carry timeless messages that are not forgotten. I started writing cycle stories to explain the process of cycles to the children I taught. The stories give children a place to collect new information about abstract ideas.

What kinds of information do we want children to collect? Educators used to think we could feed facts to students and grade them on how many they remembered. However, now we know that students learn better by experimenting and finding information themselves. Gael Keyes, my principal and mentor, showed me how to teach children to inquire. Asking questions, "inquiry," has always been the most basic force in the study of science. Therefore, each cycle in this book has an inquiry question that will, hopefully, spark many more questions in your students' minds. Turn to page 11 for a generic "Inquiry Journal Form." Copy it to use in every cycle workshop. The most beneficial aspect of inquiry learning I have found is the students' newfound ability to make and understand visual representations of data like charts and graphs.

Teachers have also learned that students learn differently. Therefore, each cycle has activities that will appeal to a variety of students. Students will find opportunities to write, act, read, listen, and create visual models. Students will have the opportunity to collect information in the mode that makes the most sense to them.

Cycle Workshops Offer:
- *Think About It! Questions*
- *Story Activity*
- *Inquiry Activity*
- *Arts Activity*

The Learning Cycle

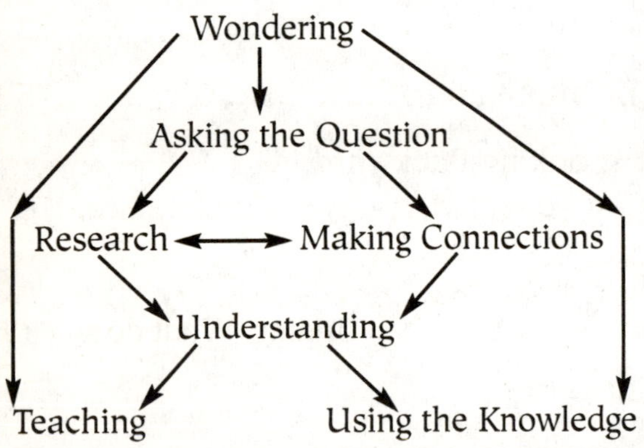

One of the final activities in the book has students assess their work. This portfolio assessment is based upon a "Learning Cycle." By choosing pieces of their own work, students will illustrate their journeys through the Learning Cycle. Let your students know from the first day that they must think about their learning and keep their work for self-assessment purposes. Before studying cycles, send home the "Letter to Parent." These questions will show that students have taken responsibility for their own learning.

Plays are a terrific tool to make stories come alive. A wonderful final project to tie learning about each cycle together would be to produce a cycle play. In the following pages novice directors will find simple steps for turning a story into a play. Give yourself permission to keep it simple. Elaborate costumes and sets are not necessary for success.

In each cycle section of this book there are three types of pages: teacher pages with mini-lessons, student pages (indicated by the pencil and magnifying glass icon), and the story (indicated by the book icon). Use *Science Cycles* in its entirety or pick and choose among the activities and topics. There is something for everyone here.

How to Fill Out the Inquiry Journal Form

When students are filling out an "Inquiry Journal" (page 11), they will always find the same parts. First, students must ask questions. They must choose their best question and record it on the first page. Next, they will fill out the "Present Knowledge" section with information already learned. Students must know that they need not worry about whether or not it will help them find an answer or help design their experiment at this point.

Now students make a prediction that is based upon what they know, an educated guess. Why? Because it makes students think critically about what is known and what is being asked.

The design of the student project is the hardest part. It is very important to fill out as much information as possible in this section. Someone who does not know the student or the subject matter should be able to pick up an individual journal and do the same experiment, exactly as it was done by the student. If it is not clear how things were measured or what or where the experiment took place, then the results are meaningless. When the design is completed, students should answer these questions:

1. Does my design really answer the question I asked or does it answer a different question?

2. In an experiment, we measure the change in the dependent variable as we keep an eye on the independent variable. For example, in the sunlight experiment we are measuring the change in area of the reflection, the dependent variable. We expect it to change as time changes. Time is the independent variable. In a good experiment, there is only one independent variable. What else might cause the area of sunlight to change? The position of the mirror. Therefore you must make sure that the position never changes.

"Data Collection" is most often in a table format. Make sure that columns and rows are labeled so that anyone reading this experiment would know what was being measured.

Making a visual representation of data is extremely important. Think about the differences between line graphs, pie charts, and bar graphs. Students should choose the best representation for their data. This makes data easy to understand.

The "Final Analysis" is where students think about the significance of their finding. What does it all mean? What new questions have come up during this experiment?

Story Theater: How to Turn a Story into a Play

General Instructions

Choose an Appropriate Audience

Have students perform in front of children who are younger than they are. Parents also make a great audience.

Choose the Type of Drama

Decide what kind of drama will work best for each story. There are several ways to bring these stories to life:

- Puppetry—You can make a quick puppet theater by turning a long table on its side and having puppeteers hide behind it. "Mole Moon" would be a good puppet show.
- Story Theater—A narrator tells the narrative parts of the story while students act out the story and say their lines.
- Readers' Theater—Actors sit on chairs in front of the audience and read their parts.

How to Do Story Theater

- As you study the story together ask the students to read parts. This is an opportunity to become familiar with the plot and the characters in the story. Whenever a new character is introduced, write the name of the character on the board. Most stories have narration, explanations of setting, thoughts, or feelings. All the stories need a narrator to read these parts except "Looking Backward, Thinking Forward." Also, during the reading, cross out any unnecessary words like, "she said and laughed." The audience will see and hear her saying the words and laughing; they do not need to have the narrator to tell them.
- After reading the story, discuss the personality of each character. Ask, "How do you think this character would move?" "How would this character eat?" "How would this character wake up in the morning?" Encourage all of your students to practice how different characters are affected by location. What would they do if they found themselves in a shopping mall, in a space station, or on the losing team at the Super Bowl? By acting together, students will be less intimidated and they learn from watching each other.
- Assign parts. If you are doing all or most of the plays, everyone will probably get several parts. Try to make sure each person gets at least one part that he or she desired.

- Blocking is extremely important when producing story theater and puppet shows. You want to make sure there is action on the stage. Try to avoid having superfluous characters standing around and doing nothing. Remind your students never to stand on stage with their backs to the audience. It is not only an unpleasant view of the actor, but any lines spoken facing away from the audience will not be heard or understood.
- Make sure that your students speak loudly. Students will often feel as though they are yelling yet nobody behind the second row can hear a single word. Arrange to have students stand in the back of the room during rehearsal and raise their arms when they cannot hear the actors. Actors can respond by speaking more loudly at this signal.
- Most audience members prefer good acting to elaborate costumes. Your actors may prefer to wear costumes because they help create a feeling of make-believe. This takes pressure off the actors to draw the audience into the play. Creating costumes can be expensive and time consuming. A compromise is to have your actors wear blue jeans and solid-colored T-shirts, adding a hat, a tail, or ears to each to make a costume whenever necessary. Students playing animals can convey the characteristics of the animal with movements and voice. Avoid masks that will cover the actors' mouths. The audience needs to hear the actors' words and see their facial expressions.
- Sets are not necessary. You may want to have blocks or solid chairs for students to stand or sit on. Remember the narrator will be describing much of the setting.

Play-by-Play Advice

Valued Things—There are only five characters in this play: the mother, the son, Connor, Sam, and the woman who gives Jack the bugs. However, you can enlist as many students as you like to play the army of ladybug larvae and aphids. Costumes could be as simple as a piece of green fabric over a student's back to identify an aphid or an army helmet to identify the ladybug larvae. Make army helmets by putting some elastic through the rim of a paper bowl. Paint the bowls appropriately. When the aphids die, they can go to the side stage, put on their helmets, and return as ladybug larvae. During the cocoon phase, the larvae can curl into balls covering themselves with black cloths previously hidden under their shirts. To reappear as ladybugs, actors slip off the black cloths, flip them over to show the other sides (red with black spots), and tie them loosely around their necks. To make the capes, staple or glue black spots onto red fabric. Then staple each red and black-spotted cape to a piece of black fabric. Staple on strings to tie or glue on interlocking pieces for fastenings.

Mole Moon—You will need five students to play the moon. Each student must have a cutout circle about 2 1/2 ft. (.8 m) in diameter. Paint the circles with bright yellow and dark blue or black paint. The full moon will be solid yellow, the crescent moon will have a yellow crescent filled in with blue, the half moon will be half yellow/half blue, and the gibbous moon will have a crescent of blue, the rest painted yellow. The new moon will be a solid black circle.

As the play opens, the narrator moves to stand behind each character as he or she is introduced, laying a hand on the shoulder of each character saying, "Mother Mole asked her babies . . ." or "stated First, the first born . . ." This will inform the audience who these characters are. To show that the moles are going outside, the student carrying the moon should pass in front of the moles as they watch. At the end of the play, the moles are all gathered to watch the crescent moon as it reappears and crosses the stage when the narrator says, " . . . they caught a glimpse . . ." When the narrator says, "And grew . . .," the half moon crosses the stage. As the narrator says, "And grew. . ." the gibbous moon crosses the stage. As the narrator says, ". . . until finally after two weeks, it was a bright, peaceful, round ball again" the full moon returns to cross the stage.

Paul Morano—The narrator's role in this play is crucial. Actors' lines are few, but their actions are very important. In the beginning of the play, have the students line pantomime bossing Peter around. As they leave, they reveal the car with Peter's mother and father inside waiting for him. The problem of showing a "car" and a "volcano" can be solved by painting a car on one side of a piece of cardboard and a volcano on the other. Then as Peter's mother and father drive away, Peter flips the car over and becomes a volcano. Have pieces of confetti and crepe paper streamers that Peter can throw at the audience to liven up the show. Great-Aunt Martha can sweep it all up, shaking her head. When the Tyrannosaurus dies, have students bring in pieces of fabric or painted paper to represent the layers of mud, clams, and fish bones to cover the ground. A student with a sign that reads "Pressure" can push down on the layers. At the end of the play, actors come on stage and become statues that Peter can study or carve. Peter finishes the play, stepping forward to explain his work.

Where Is the Rain?—In this play, actors will become agents of nature like the wind, ocean, mountain, sky, and river. A piece of cloth draped over each actor's shoulders is enough to transform them.

If possible, choose a small, light actor to play Iris so that she can be whisked up by the wind and swirled around the stage to meet the sky.

The mountain's snowy white peak can be a piece of white fabric when he or she is wearing it. When the mountain hands it to the river, the prop should be flipped over to become a blue, watery color. This can be easily done by stapling together two pieces of fabric, one white and one blue.

To increase interest in the long walk that Iris must take, ask several students to change from thirsty lizards (lying on the ground) to shriveled bushes (crouching) to parched people (standing with hands out to Iris).

Two students should hold a blue piece of fabric in front of the ocean and make gentle waves with it while the ocean is talking. When the ocean's waves crash in return to what Iris says, the students make larger waves with the blue fabric.

Eating Sunlight—In this play it is important to remember that it may be difficult for the audience to see actors on the floor. You will have to think about how the animals like Snake, Toad, and Grasshopper will be seen, unless your actors are playing on a raised stage. One possibility is to have a table that the characters climb up on before they speak. It is important that they use animal movements to get up and down. To put Owl in the tree, ask a taller student to play the apple tree. Then Owl can stand on a chair or block and peek through the boughs (arms) of the tree.

Looking Backward, Thinking Forward—The main setting of this story is in a classroom but there are flashbacks to the other side of the stage area as Mr. Miner recalls his life. There is no narrator, but Mr. Miner has to appear as a baby, a child, a young man, and an old man. Since the old Mr. Miner never leaves the stage, you need someone else to play Harry as a child and as the young man. Whenever a flashback occurs, the classroom actors should all freeze and listen. As soon as the flashback actors are finished performing, they should exit the stage and leave it empty for the next flashback.

Dear Parent/Guardian,

In our classroom we will be studying cycles. The emphasis of this unit will be on asking solid questions and finding complete and understandable answers. A good way to start is by finding out what is already known about a topic and deciding what you would like to learn from the unit.

Some of the science cycles that we may study are:

Sun Cycle	Water Cycle
Moon Phases	Carbon Cycle
Rock Cycle	Recycling

Having read an outline of the unit, make sure that your child thinks of three questions about cycles. At the end of this unit of study, your child will be responsible for trying to find answers to these questions. Write the questions below.

1. ____

2. ____

3. ____

Thank you for your help and support,

Name ______________________________

Inquiry Journal

Best Question: ______________________________

____

____

Present Knowledge:

Use this box to help organize your knowledge.

Written Present Knowledge: ______________________________

____

____

____

____

____

Name ________________________________

Inquiry Journal Page 2

Prediction: ________________________________

____

____

Design: ________________________________

____

____

____

____

____

____

____

____

(Draw a model of your design if appropriate.)

Name ______________________________

Inquiry Journal Page 3

Data Collection: (Draw table or other organized form.)

Notes About Data (if needed): ____________________

____

____

____

____

____

____

____

____

Name ______________________________

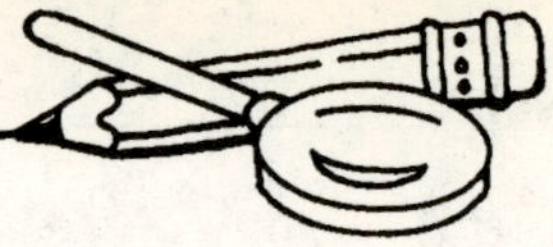

Inquiry Journal Page 4

Data Presentation: (Attach graph, chart, or diagram here.)

Final Analysis/Conclusion: ____________

____

____

____

____

____

____

____

____

____

The Sun Cycle

Reading the Story

(large group or individual)

Materials:

- Copies of "Valued Things" if reading independently

Either read "Valued Things" aloud to the class or have the students read silently. When finished, discuss the plot and themes found in the story. Explore the topic of values with your students.

Mini-Lesson: Valuing Others

Ask students to think about people they value. These people should be members of their families or community. Invite each student to make a web about the person they choose by putting the person's name in the middle of a blank page. On this web, have each student list good attributes about the person. Continue branching from the name by writing a specific incident to illustrate that attribute. Discuss the format of a personal letter with the students. Ask the students to write letters to the people they have chosen as a valued part of their lives. Use the web to make sure that each letter contains specific examples of why this person is valued in their lives.

Inquiry Journal Activity

(small group or individual)

Materials:

- Inquiry Journal form, one per student
- Graph paper, one per student
- Tape
- Rectangular or square mirrors, one for each experiment (Many glass companies will donate hand-size mirrors for free. Tape the edges of broken mirrors to prevent injuries.)
- A large box for each experiment
- White poster board to cover the back and sides of the box (This allow students to get more readings by widening the backdrop.)
- Paper to cover the cardboard backing (optional)
- Markers to lightly color shapes on the mirrors (optional)

Teaching Tips:

The best way to introduce this activity is to insert a pencil into a piece of heavy cardboard. Students may draw the position of the shadow on the cardboard every hour as shown.

Teaching Tips:

To extend this activity, use the data the students gathered and make a sundial out of clay. Fire and glaze, so it will last outside for years. Set the dial in a sunny location.

Have a class discussion about the apparent path of the sun across the sky. Discuss how the sun's "path" changes throughout the year. Also talk about how Earth revolves around the sun.

Students may not know about how light is reflected by mirrors, (the angle of reflection is equal to the angle of incidence). This may be a good time to give them mirrors and flashlights or just go outside and use the sunlight to let them experiment. Remind the students never to reflect light into others' eyes.

General Suggestions:

For this inquiry, the "Prediction" may be a picture of what the student expects to see. Scientists often draw diagrams to explain their thinking. Make sure that students understand that all diagrams must be labeled so that others can interpret their diagrams.

Allow students to decide on their experiment. Be careful about where students set up experiments. Discuss what direction they will want their mirrors to face to get a clear view of the sun. Watch how shadows shift during the day before selecting a location so it won't fall into shadows in the afternoon.

Before setting up the equipment, make this experiment more intriguing by asking students to color geometric shapes onto their mirrors using markers. Now their projects are light sculptures. Leave blank spaces around the outside of the mirror and between the shapes to define the edges. It is interesting to see how shapes change as the day progresses.

Try to make time in the schedule for students to go outside and trace the shape of the reflected light every hour. Set up your day in advance to allow this to happen. Make sure students have a table or chart ready to record their data before going out to take the first reading.

For "Data Presentation" students should make a line or bar graph that shows how the area of the shape changes. Have students create their representations on graph paper. Make sure they understand how to title and label graphs. Also have them include a key if needed.

Hopefully, in the "Final Analysis/Conclusion," students will make the connection that it is generally hotter and the sun is brighter when it hits Earth directly at high noon. This is also true of the sun's energy during the summer solstice when the sun's rays hit us most directly. This generalization will lead students into the next activity—a seasons mural.

Making a Seasons Mural

(small group)

Materials:

- A large piece of paper in the shape of a circle, divided as shown
- Construction paper, scissors, glue, paints
- "Your Community and the Cycle of Seasons" worksheet, one per person or group

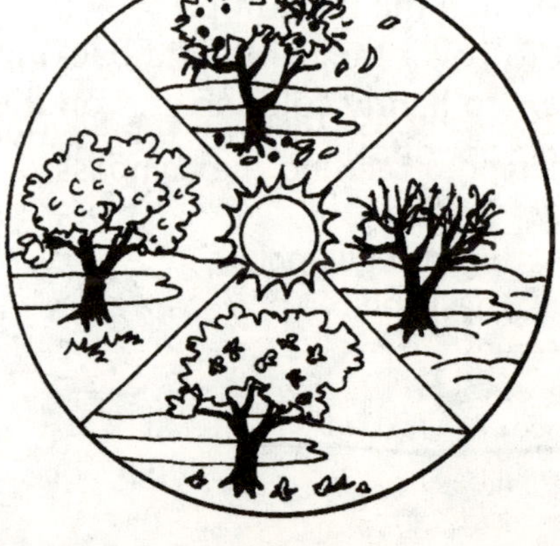

Decide if there will be groups of four students to make one mural or if one-fourth of the students will study each season. Assign a season to each group. By filling out "Your Community and the Cycle of Seasons" worksheet students will research how they are affected by the changes in the angle of the sun's light. Before students start cutting paper to make plants and animals to glue onto the mural, give each group a cutout figure of a person. This can be used to consistently scale everything throughout the mural.

When the students have finished the mural, ask them to write a report about the season they have just illustrated. In this report they will discuss how the changing angle of light in their community affects them.

Mini-Lesson: Scales and Ratios

Given a figure of a person, it is easy to determine the size of everything else in a mural. Perhaps the object would come up to a person's knees or it would be the size of a person's hand. Another way that to find how big to make things is to use ratios. If a five-foot (1.5 m) person were only 10 inches high, how high would a dog be? Start with the scale given.

$$\frac{\text{60 in. (1.5 m) (person)}}{\text{10 in. (25 cm) (picture)}} = \frac{\text{24 in. (61 cm) (dog)}}{\text{X inches (picture)}}$$

Doing the math by cross multiply and solving, students will find that the dog will be 4 inches tall. Even if the student's math skills do not allow them to do this math, they may see that the picture is always six times smaller than the real thing.

Mini-Lesson: Geometric Constructions of Perpendicular Lines

To make the equator and latitude lines on the diagram of Earth, use geometric constructions. Begin by explaining to the students how to use a compass. To make an equator:

1. Start with the compass point on the North Pole.
2. Open the compass wider than halfway across Earth, less than the diameter of Earth.
3. Draw an arc.
4. Move the compass point to the South Pole without opening or closing the compass.
5. Draw an arc.
6. There should be two points where the arcs cross. Connect these points and it will have bisect the diameter of the Earth with a perpendicular line. Label this line "Equator."

Tilting Earth Model

Materials:

- Paper, pencil, markers, scissors, glue

Ask each student to draw a circle to represent Earth on the paper. To locate the axis, first draw a vertical line slightly larger than the circle. Show the students how to make a line from the bottom of the vertical that forms an angle to the right of the vertical line. Using a protractor, draw a thick line coming from the bottom of the line out at approximately 235°. This new line is the Earth's axis. Ask students to cut out their Earths and label the North and South Poles. Have the students place their Earth's around the mural showing that the sun is hitting Earth more directly during the summer and less directly during winter.

Writing a Song about the Seasons

(large group and small group)

Materials:

- Copy the lyrics for "Twinkle, Twinkle Little Star" on poster board

Before writing lyrics, students should understand rhythm, meter, and rhyme. This can be done using the lyrics for "Twinkle, Twinkle Little Star" and examining the rhyming pattern and the pattern of accented syllables. Have students write lyrics for a new first verse to the song extolling the virtues of the sun. The title of this song can be "Twinkle, Twinkle Little Sun."

Mini-Lesson: Stressed and Unstressed Syllables in Poetry

A meter in poetry is the rhythmic repetition of a pattern of stressed and unstressed syllables. Clap out the first line of "Twinkle, Twinkle Little Star." It is stressed-unstressed, stressed-unstressed, stressed-unstressed-stressed. Ask students to find names of classmates that will make the same pattern.

Research Resources

Books

Asimov, Isaac. *Why Do We Have Different Seasons.* Milwaukee: Gareth Stevens Children's Books, 1991.
This is a good reference for why the weather changes as our tilted Earth moves around the sun.

Arnosky, Jim. *Crinkleroot's Nature Almanac.* New York: Simon & Schuster Books for Young Readers, 1999.
This has students discover the changing face of nature through the year with Crinkleroot.

Clark, Ann Nolan. *Circle of Seasons.* New York: Farrar, Straus & Giroux, 1970.
This book describes how Pueblo Indians celebrate and participate in the changing seasons.

Hall, Donald. *Ox-cart Man.* New York: Viking Press, 1979.
This story may encourage students to think about how their community is affected by the changing seasons.

Simon, Seymour. *Earth, Our Planet in Space.* New York: Four Winds Press, 1984.
This provides a simple explanation of why seasons change.

World Wide Web Sites

http://www.astro.queensu.ca/~hanes/p014/Notes/Topic_006.html
This site, one page from the Queens University Astronomy Research Group, has a wonderful description of the reasons for our seasons.

http://www.usatoday.com/weather/wsunrise.htm
You can find links to find any data you might want about the sun, moon, stars, and time. It also links to weather sites.

http://starchild.gsfc.nasa.gov/docs/StarChild/StarChild.html
This is a great place to let children explore. There are wonderful links to varied and rich astronomy sites.

Name ____

The Sun Cycle

Think About It!

How does energy from the sun vary during a day? During a year? How does life on Earth respond to the cycle of more direct sunlight (energy) and less direct sunlight (energy)?

Inquiry Journal Activity

Best Question: "How does the movement of the sun across the sky change the amount of direct sunlight we receive?"

Present Knowledge: Fill in the section by thinking about the path of the sun across the sky. What changes in sunlight and heat do you notice during the day? How does it change as the year progresses? In this inquiry, you will be watching sunlight reflected by a stationary (unmoving) mirror. Add to your "Present Knowledge" section anything you know about how light is reflected by a mirror.

Prediction: Based upon what you know, what do you predict you will find? Your "Prediction" may be a diagram of the changes you think you will see.

Design: In order to watch how the sunlight hitting the Earth (your mirror) changes during a day, you will need a small mirror and a box onto which to reflect the sunlight. Make sure that you choose a place to set up your experiment that will be in sunlight all day. How will the surrounding buildings' shadows move during the day? Will their shadows cover up your experiment? Think about how you will record your data. You should trace around the reflected sunlight but will that be enough. Do you also want to measure the length and width of the reflected sunlight? Don't forget to write down the time of each reading.

Data Collection: As you complete the experiment, record your data on a table you have drawn on your "Inquiry Journal".

Data Presentation: Use the data you collected to make a line or bar graph. Since our question asked how the amount of direct sunlight changed during the day, you will want to graph the area of the reflected sunlight. If the shape of the sunlight is not rectangular, estimate it. Time, (the independent variable) will be on the *x* axis. Area of reflected sunlight, (the dependent variable) will be the *y* axis. Why is time the independent variable? Is there anything you can do to change it?

Final Analysis/Conclusion: Look at the graph you have drawn and see if it looks like what you expected. Is the line straight or curved? Why? To make meaning from this data, you

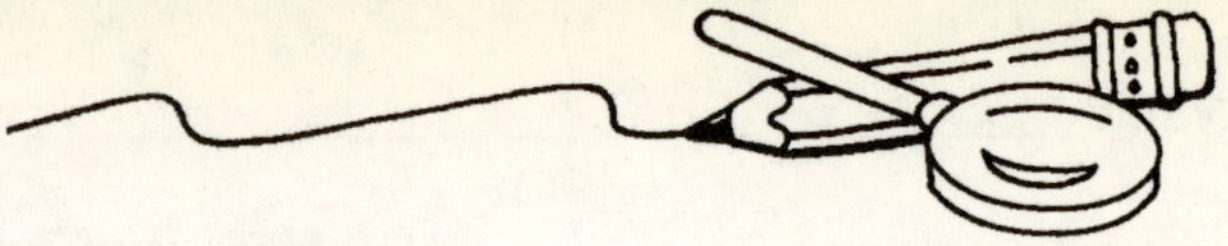

will have to think about this: The amount of sunlight (energy) being sent to Earth from the sun remained the same. What then, does it mean if the same amount of energy is spread out over a larger area?

New Investigation: Predict what you might notice during times when the area of reflected light was largest. When is this time? What would you notice when the area was smallest? At what time during the year would this happen? How might this information teach you about the changing energy Earth receives as it orbits the sun?

Making a Seasons Mural

During the year, the sun's rays become more or less focused. This is caused by the Earth's tilt. We recognize these changes as seasons. Using the circular paper provided by your teacher, make a mural of seasons. You will be assigned a season to research. To learn about seasonal changes in your community contact your Chamber of Commerce. Interview friends and members of your family and community. Read old newspapers. Fill out the "Your Community and the Cycle of Seasons" worksheet. Use this information to make a complete picture of your community during the season you were assigned.

Write a report: After your mural section is complete, use it to write a report describing how the changing angles of light cause changes in your community.

- **Paragraph 1:** Explain the position of our tilted Earth in comparison to the sun during your chosen season.
- **Paragraph 2:** Discuss how animals adapt to the season.
- **Paragraph 3:** Explain how plants and farming are affected.
- **Paragraph 4:** Address the people in your community and how they are influenced by the seasons.
- **Paragraph 5:** Sum up all of the information in the report to make a conclusion about how your community responds to the affects of our Earth's tilt.

Tilting Earth Model

Make a tilting Earth by following instructions provided by your teacher. Upon completion, tape your Earth around the mural to illustrate how the tilt of the Earth causes the seasons to change.

Writing a Song about the Seasons

With other students, change the lyrics of "Twinkle, Twinkle Little Star" into a song about our favorite star, the sun. As a large group, write one verse explaining how we rely upon the sun. Then the group of students who worked on the summer section of the mural will compose lyrics for a summer verse. Students who worked on the fall section of the mural will compose fall lyrics, and the students who completed the winter and spring sections will complete those lyrics.

Name ________________________________

Your Community and the Cycle of Seasons

Before working on the mural, answer these questions completely. You may want to call your Chamber of Commerce and the Chamber of Commerce in nearby cities or resorts. Ask for pamphlets describing the region and its attractions. Search the Internet for information describing local events, festivals, or natural resources.

1. What animal behaviors are observed during this season?

2. How are plants (domestic and native) reacting to the season?

3. Are there any festivals (music, ethnic, athletic or any other) or celebrations that occur during this season?

4. Why would tourists be attracted to your area during this season?

5. What do kids do for fun during this season?

6. What foods are eaten during this season?

7. What clothing do people wear during this season?

Valued Things

"You know Jack, today would be a great day to build a snow fort." Jack's mother said. Near the window, a sad Jack watched the snowfall. He said nothing. Jack's mom walked behind him and put her hand on his shoulder. "I know you are sad because Maggie moved away. You were best friends for a long time, but maybe if you go out you will meet some new friends."

Jack shrugged his shoulders. "I did see a couple of new kids down the street. They were outside yesterday when Maggie left." Jack put on his snow clothes and went to find his shovel.

"B-r-r-r-r," Jack shivered. "I'll walk down the street to see if the kids are outside." Being too shy to knock on the door, Jack walked up and down the street until his nose was cold. He was about to turn toward his house when he heard a screech. An old VW bug spun out of control and landed in a snow bank.

The door opened and out stepped an upset woman with long brown hair and bug-eyed glasses. The two kids Jack had seen the other day were in the back seat. "Oh!" she exclaimed. "And I almost made it home." The two boys jumped out of the car to investigate the stuck car.

Jack was glad to help. "I have a shovel," he said. "Let me dig you out."

"Well, you must be our guardian angel," the woman said. "If you dig and Connor and Sam push while I drive, we'll be out in no time!"

Jack dug around the tires, and Connor and Sam gave the car a mighty push. The car leapt out of the snow bank. Motioning for them to follow, the woman drove into the driveway.

As they started up the driveway Connor

said, "Thanks for your help. We are staying with my aunt because we just moved from Florida. Our parents are down the street looking at a house that is for sale."

"I bet it's my best friend Maggie's old house. She just moved too! Do you like building snow forts?" Jack asked.

"They would love to," their aunt said. "But not today because we have to go in to eat lunch. Maybe tomorrow."

"Tomorrow would be great!" Jack beamed. Connor and Sam smiled and promised to walk down in the morning.

"Here, guardian angel Jack," their aunt leaned down and scooped up some dirt and a dead plant from her garden. "Here is a special gift. Just bury this in your garden. In the spring, you'll find a treasure."

Jack wasn't sure whether to believe her or not. It didn't really matter since he had made two new friends. Jack smiled as he walked home.

At home, his mother met him at the door saying, "Did you finish your snow fort?"

"Oh, mom," he said excitedly, "I forgot about the fort, but I made two new friends and their aunt gave me this for helping her out of a snow bank!"

"Well, that is interesting." Jack's mom looked quizzically at the shriveled plant. "Why don't you throw it out in the garden, and we'll plant it later. Then get washed up for supper."

Jack threw the dirt and plant out the back door. The wind caught them and blew them into a crack in the garden wall. He went in to dinner and forgot all about his treasure.

Jack, Connor, and Sam played for the many more snowy months that followed. But winter always turns to spring, and so the soil in the withered garden behind Jack's house warmed. He remembered Connor and Sam's aunt and the gift she had given him. Somehow, he found himself drawn to the garden. He thought that it couldn't hurt to water the sad-looking roses, so he did.

His watering was rewarded and little leaf buds sprouted. When the shiny, green leaves opened, the trouble began. Aphids. At first, just a few nibbled on his leaves. Then there were hundreds, then millions.

"It's always the same, my projects never work out." Jack hung his head. Looking at the dirt, he saw something surprising. There was a black fuzzy insect that looked like a six-legged alligator with orange spots. Behind it there was another and another and another. They marched in a line from a crack between two bricks in the wall. He dug his finger into the crack and saw the leaves and dirt that had been given to him in the winter.

"Oh, this is terrific!" he exclaimed. "This army of bugs is my treasure! I love bugs!"

As he watched, the bugs did something he hadn't expected. They didn't eat the plants at all. They started munching on the aphids! In a month's time they'd eaten zillions of aphids.

Jack's garden grew. The flowers were gorgeous on the day that his bug army curled up and appeared to die. Four days passed and each one brought more aphids. Jack was very disappointed.

"I'm sorry that the aphids are coming back." His mother said when he told her.

Then, a miracle happened. As Jack clipped the last of his velvety roses, he noticed the curled up bodies of his bug army were splitting open. They weren't dead; they were cocoons. Jack watched in awe to see little orange beetles with black spots crawl out.

"Ladybugs!" Jack shouted. "Sam and Connor's aunt gave me ladybugs. I can thank her with these roses!" He tried to find her house but everything looked different now. No snow. Green bushes and trees hid the houses. "I'll have to go next door and ask Connor and Sam," he decided and started for home.

"Kid," a gruff voice said, "I'll give you 10 bucks for those flowers. My wife loves roses."

"Sure," Jack said, handing them over as he accepted the money from the old man. Jack sniffed the crisp 10-dollar bill and held it up to the sun in disbelief. "Green for my greens," he laughed. Jack smiled and hatched a plan right there. For the rest of the summer, he sold his flowers. When Jack picked his last rose of the summer, he gathered all the money he had earned and wrapped it in brown paper, giving it to his mother.

Jack's mother put the rose in an old pickle jar to always remember his kindness. "You need to save your money Jack. Maybe some day you will open your own flower shop!"

The Moon Cycle

Reading the Story

(large group or individual)

Materials:

- Construction paper, glue, scissors
- Copies of "Mole Moon" if reading independently
- "Phases of the Moon" worksheet, one per student
- Reference materials about the moon

The first step to introducing the moon is to ask students to draw a picture of the moon. Place all of the moon pictures on the board and look at them. How are the pictures alike? How are they different? Did most of the students use black or dark blue backgrounds? Do any of the pictures show the moon up during the daytime? Has anyone noticed that the moon is sometimes up during the daytime? Either read "Mole Moon" aloud or have students to read the story silently. Afterwards, have students complete the "Phases of the Moon" worksheet. Discuss the worksheet and then encourage the students to think of questions they would like to answer about the moon.

Inquiry Journal Activity

(small group or individual)

Materials:

- Inquiry Journal form, one per student
- Moonrise and moonset times, located in Farmer's Almanac or at Web site <http://www.usatoday.com/weather/wsunrise.htm>
- Copies of "Moonrise and Moonset Times" worksheet
- Pencils, markers, scissors, glue

General Suggestions:

This inquiry activity will address the "Best Questions": "Why does the moon sometimes appear in the afternoon? Does the shape of the moon affect when it appears in the sky?"

A week or so before beginning this unit, ask students to go outside and watch the moon at the same time each day and night. Is the moon up when they go to bed? Is it up when they go to school? What shape is the moon?

Their "Present Knowledge" should include moon shapes and everything else they have learned about the moon. Make sure they include all the information from their "Phases of the Moon" worksheet. Now ask your students to write their "Prediction" about why the moon might be up in the sky in the middle of the afternoon. Is a daytime moon always the same shape?

The "Design" of this experiment calls for students to color in circle graphs for a month of moonrises and moonsets (see page 33). Tape the 32 graphs into a long strip to represent the month. Ask them to date the graphs and color in the moonrise and moonset times as shown in the example.

The shape of the moon is important to this inquiry so ask the students to find the date when the moon was full. To complete the "Moonrise and Moonset Times" worksheet, cut out the full moon from the "Phases of the Moon" worksheet and paste it in the center circle of the graph as shown in the diagram. Then cut out the next moon shape and paste it in the graph representing the next day after the full moon phase. Continue until the days following the full moon are filled. Now ask the students to cut out the moon from the day before the full moon and paste it onto the graph of the day before the full moon. Have them work their way back through the month until they run out of moon shapes to cut out. Some days will not have moon shapes on months with more than 28 days.

Mini-Lesson: Why the Moon Changes Shape

Create a model of the sun-Earth-moon system by asking one student (the sun), to hold a lamp, another to hold a globe, and the third to hold a tennis ball (the moon). Ask the moon to travel around Earth. How does this information help explain their data? The next time the moon orbits around Earth, ask the student to stop at the full moon phase. Have them look carefully at the globe. Ask who can see the moon now? What time of day is it for people who can see the moon now? Have the moon continue around Earth. Stop the moon when it is the new moon phase. Ask who can see the moon now? What time is it for the people who see the moon? Expect to do this demonstration several times until all students understand the concept.

The last piece of information students will need to put on their graphs is the time at which the sun rises and sets each day. This will not vary much during the month they study but could be considerably different when comparing, for example, the months of June through December.

Ask students to draw a 24-hour time line in their "Data Presentation" section. Place a full moon shape at the time their full moon rose. Draw in a new moon shape when the new moon rose. Do the same with the quarter moons and gibbous moons. Mark in the approximate sunrise and sunset times for this month.

When all groups have finished, arrange the long graphs next to each other with full moons lined up. Ask your students to look for patterns in the moon shapes and the rising or setting of the moon. Students should write down any observations made in the "Final Analysis/Conclusion" section on the form. Students should see that the full moon rises at sunset, the new moon rises at sunrise, etc.

Students should try to relate the shape of the moon to the time that it rises. What other questions have come up? Could they understand the occurrence of a lunar eclipse better using the information they have gathered? Do people living on the opposite side of Earth see the same moon shapes we do? How long is a lunar day (sunrise to sunset)? An old wives tale says more babies are born during the full moon phase. Is it true? Use the information from the web site to decide if this is true for the families in your class: <http://seds.lpl.arizona.edu/nineplantets/nineplanets/luna.html>

Mini-Lesson: How to Write a Newspaper Article

When writing a newspaper article, the traditional approach was to answer the five W's and an H (who, what, when, where, why and how) as quickly as possible. Details would follow later. Draw a triangle on the board and explain that information should flow in that shape: the most important information is provided at the beginning of the article, less important information is next, and the information at the end is be interesting but not as important to understanding the event. The reason for this was to allow busy readers to move on quickly to another topic when their questions were answered. Have students compare and contrast this writing style to that of a good story.

Exploring the Moon Activity

(small group)

Have each group of students choose an important date for exploration of the moon on page 30. The students' job is to create, using their imagination, a newspaper from that date in history. Before beginning, they must bring in samples of the different parts of a newspaper that they will create: a front-page article, a letter to the editor, an editorial cartoon, and advertisements. Once the students have brought in samples for discussion, they can begin researching their date. Since listening is so important to any unit of study, students must take notes while one student reads aloud from source material. Afterwards, students share what they felt was important from the reading. Now they must ask themselves as a group, what do they still need to know to create the newspaper? How can they find that information? When they feel they have enough information, they should assign jobs and start writing the articles.

Puppet Show Activity

(small group)

Materials:

- Craft materials for puppets and set
- Pencil, paper, scissors, glue, markers, paints

Students will write and perform a puppet show set on the moon. The characters must solve a problem that came about because someone was not listening. As the play unfolds, they should drop clues to let the audience know during which moon phase the story occurs.

Research Resources

Books

Asimov, Isaac. *How Did We Find Out About Outer Space?* New York: Walker, 1977.
This traces the history of mankind's venture into outer space.

Johnstone, Michael. *The History News: in Space.* Cambridge, Mass.: Candlewick Press, 1999.
Uses a newspaper format to look at developments that led from the ideas of Copernicus and other early scientists to moon exploration and beyond. Share this with your students or keep it as a teaching tool.

Sullivan, George. *The Day We Walked on the Moon: A Photo History of Space Exploration.* New York: Scholastic, 1990.
Traces the technological developments which have made it possible to explore outer space and walk on the moon.

Vogt, Gregory. *Apollo Moonwalks: The Amazing Lunar Missions.* Berkeley Heights, NJ: Enslow, 2000.
This book discusses the six Apollo missions that landed on the moon, describes the work performed there, what it was like to walk on the moon, and the collection of moon rocks.

Web Sites

http://seds.lpl.arizona.edu/billa/tnp/luna.html
This is Bill Arnett's moon page. There are visual images of the moon in calendar form and good information with many great links to other moon Web sites.

http://starchild.gsfc.nasa.gov/docs/StarChild/StarChild.html
This is a site for children interested in astronomy. There are many great links to other sites.

http://www.usatoday.com/weather/wsunrise.htm
This is the place to find moonrise and moonset times for research.

http://www.magna.com.au/~prfbrown/i_asimov.html
Read "The Triple Triumph of the Moon." Learn about Isaac Asimov's theory about how the moon has helped people.

http://www.lpi.usra.edu./
Provides information about moon exploration, includes a discussion about the future; provided be the Luanr and Plnaetary Institute.

http://www.solarviews.com/eng/history.htm
This shows a history of moon exploration.

Name ____________________

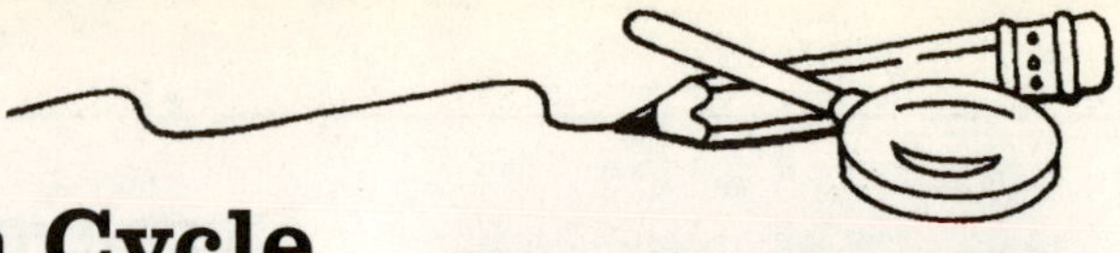

The Moon Cycle

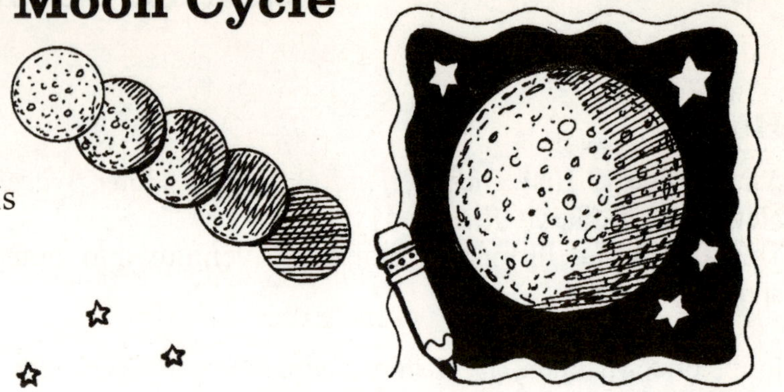

Think About It

Why does the moon change shape? Is there really a dark side of the moon?

Inquiry Journal Activity

Two questions that might have come up during your discussion about the phases of the moon are: "Why does the Moon sometimes appear in the afternoon or morning?" "Does the shape of the moon affect when it appears in the sky?"

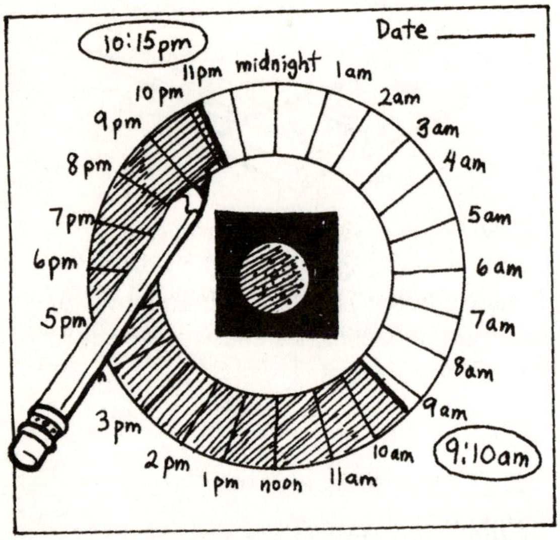

To answer these questions, select a month and fill out the sunrise/sunset, moonrise/moonset information for each day on the graphs provided. Cut out the pictures of the moon on the Moon Phases worksheet and glue them onto the graph. (If it is 15 days from the last full moon, count 15 moons from the full moon, cut it out and glue it on the graph for that day.) When everyone has completed their month of moonrise/moonset graphs, we will put them up and compare the information. Afterwards, by reading the provided books, see if you can come up with a theory to explain what you have discovered.

Moon Exploration Activity

Is there really a dark side of the moon? No, the dark side of the moon is only the side of the moon we never see from Earth. To learn about how we traveled to the moon and finally saw the hidden side, choose one of these dates and research it.

- Galileo observes the moon with a telescope—1609
- Balloon flight—1912
- Sputnik, USSR, first artificial satellite—1958
- Friendship 7, USA, John Glenn, first American in space—1962
- Ranger 7, 8, and 9, USA, sends pictures of the moon back to Earth—1964–65
- Apollo 8, USA, first manned lunar fly-around and Earth return—1968
- Apollo 11, USA, first manned lunar landing—1969
- Apollo 13, USA, explosion during translunar coast makes it a mission to survive—1970

Name ______________________________

Create a newspaper front page that might have been published on the day you researched. Before you begin, collect at least three examples of each section of a daily newspaper. After collecting important facts and interesting details, begin writing your newspaper article.

Puppet Show Activity

As a group, brainstorm ideas about the number of characters you will have in your puppet show. Have one person act as a secretary and write all of the ideas on paper. After you have your list talk about each character. Describe their personalities—what kind of person each is, his or her likes and dislikes, and other characteristics. This information may not be used in the play, but will help you write down the words and actions each might use in the show.

Next discuss the basic storyline. In this case, think about the future of people living on the moon. The group living on the moon, in the Sea of Tranquility, will experience a difficult situation because some characters refuse to listen to each other. Brainstorm the types of problems that might occur on the space station, the effects of the problems, and the ways that the different characters might deal with and solve the problems. This information will become the action of the puppet show.

Remember that each person in your group has strengths that can be used to make the puppet show a success. Some members may be creative writers, others may be able to easily see where people should stand and how the puppet show should move. Other people may be good at making simple backgrounds for the play.

As the writers put together all of the ideas that came from your discussion, remember that it must be written in a way that everyone in the group knows when and where the puppets enter and exactly what each will say. Because this lesson is part of the moon phases unit, be sure to give the audience hints about whether the events happen during the phase of the full moon, quarter moon, gibbous moon, or new moon.

When the puppet show has been written, have several rehearsals, making sure that all members of your group know their individual parts in the action.

Name ______________________________

Phases of the Moon

Draw a picture in each box showing the shape of the moon each mole described and match it to the following words:

A) new moon
B) quarter moon
C) full moon
D) crescent moon
E) gibbous moon

First's view __________	**Second's view** __________	**Third's view** __________
Fourth's view __________	**Fifth's view** __________	**Your view (moon scene)** __________

How many days does it take for the moon to complete a cycle?__________

Save this worksheet! You will paste your pictures of the moon on your moonrise/moonset graphs.

Name ______________________________

Moonrise and Moonset Times

Find out the moonrise and moonset times for each day of the month. Shade in the times on the graph. Also fill in the sunrise and sunset times each day. Cut out the corresponding picture of the moon phase and glue it in the center of the circle graph.

Date ____________

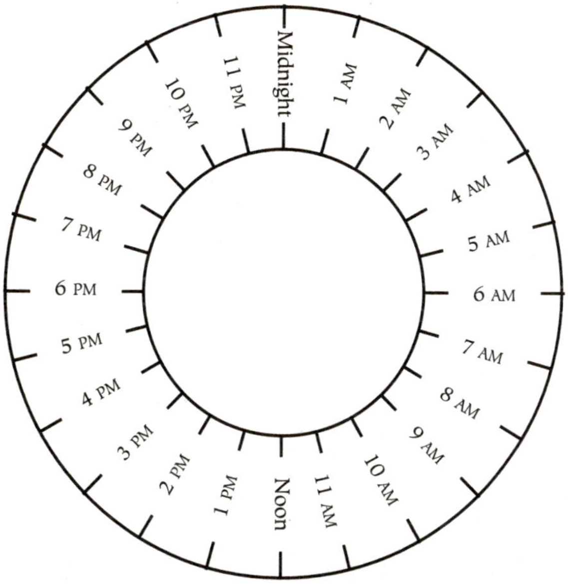

Sunrise __________ Sunset __________

Date ____________

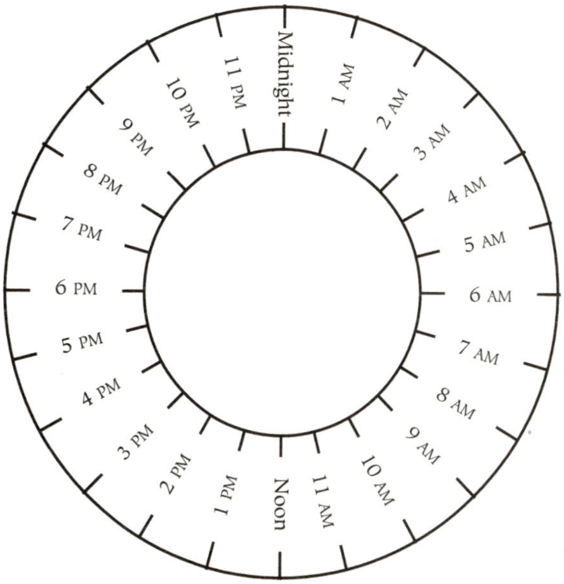

Sunrise __________ Sunset __________

Date ____________

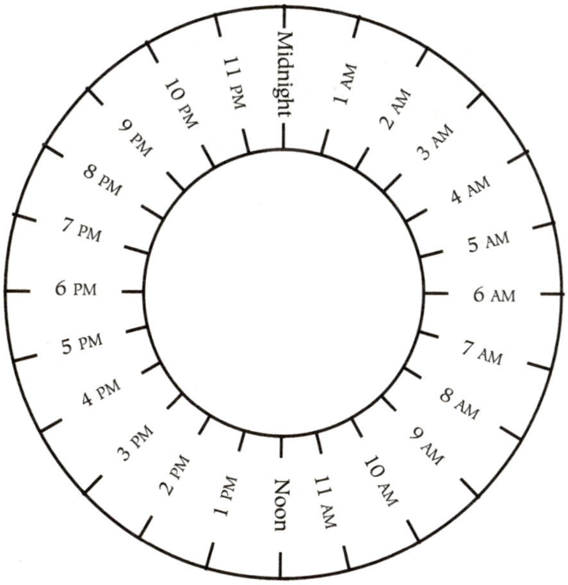

Sunrise __________ Sunset __________

Date ____________

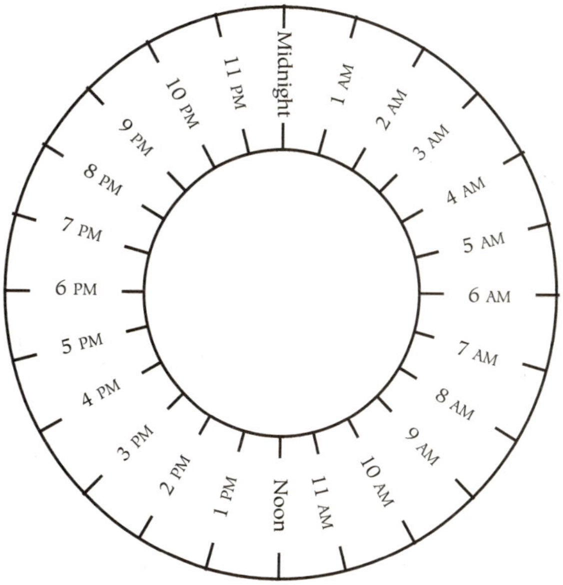

Sunrise __________ Sunset __________

Mole Moon

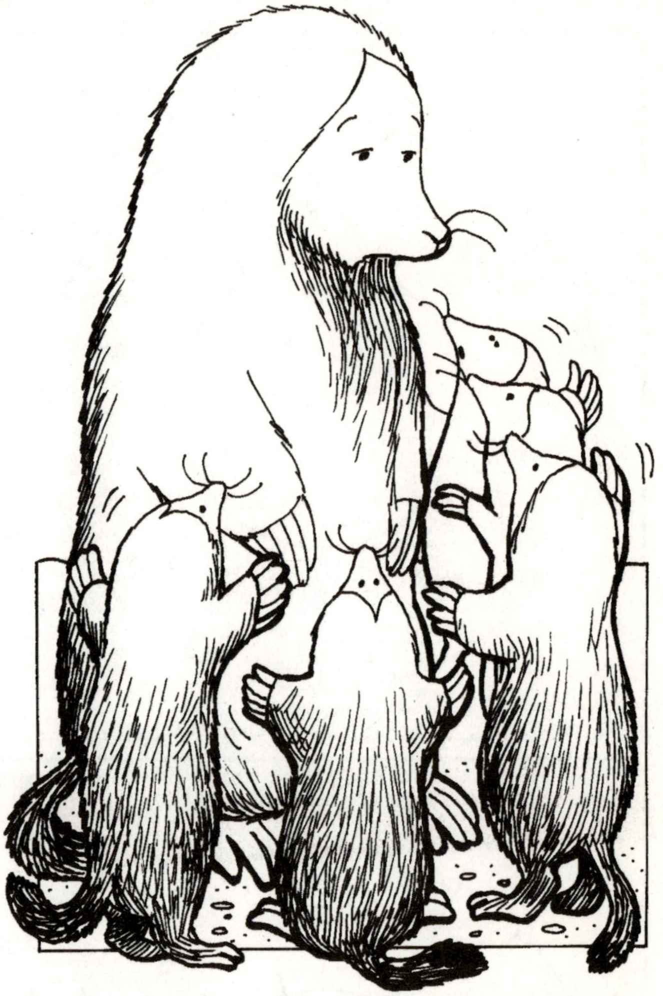

"Don't you just love to dig in the soft dirt near the shed?" Mother Mole asked her babies one Sunday night.

"The dirt is softer under the potatoes!" stated First, the firstborn of five baby moles. "No, it's softer near the peanuts," said Second, the mole baby born one minute after First had been born.

"Wrong! Anyone can tell the dirt is softer by the old rock wall," said Third, born only one minute after Second.

"I like it by the water hose because it's slimy and wet!" chimed in Fourth who was only three minutes younger than First.

"Who cares?" said Fifth. "I'm hungry."

"We care!" shouted First through Fourth all at once. Then each loudly recited his favorite digging spot and the reasons why he felt it was best. Fifth merely rubbed his tummy and recited all the foods that he wished he was eating.

"Baby moles!" said Mother Mole. "I'm going out into the moonlight to see what I can see."

The five mole babies looked at her in astonishment, then in unison, called after her, "Can I come, too?! I want to see the moonlight."

"I will take one of you with me, but only one because we need to quietly see what we can see."

Mother Mole and First strolled in the bright moonlight quietly. "Do you see the moon?" she asked.

"It's beautiful," said First, "bright and round and peaceful."

"Yes, it is. Try to remember it exactly how it is right now. Now, let's go back in."

For a few days, the mole babies got along with each other until a disagreement broke out concerning which food is the most delicious.

Mother Mole calmly said with a smile, "Oh no, what silly bickering. I am going out to walk in the moonlight." Hearing this, Second through Fifth called out, "Can I come, too?! I've never seen the moonlight. First says it is bright and round and peaceful."

Mother Mole said, "I will take only one of you with me, as we need to quietly see what we can see. Second may join me."

As they strolled, Mother Mole asked, "Do you see the moon?"

"It's marvelous," said Second, "bright and squashed on one side and peaceful."

"Yes, it is. Try to remember it exactly how it is right now. Now, let's go back in."

The mole babies tried their best not to argue again, but by Sunday they could not stop themselves from disagreeing about which game is the best to play.

"Oh my, what a lot of noise," sighed Mother Mole. "I am going out to walk in the moonlight." Hearing this, Third, Fourth and Fifth called out, 'Can I come, too? I've never seen the moonlight. It is bright and peaceful. Maybe it is round and maybe it is squashed. Please take me."

"Third needs a chance to quietly see what we can see," said Mother Mole.

"Look. It's magnificent," said Third as he looked up at the moon in amazement. "It's bright and half a circle and peaceful."

"Yes, it is. Try to remember it exactly as it is right now. Now let's go back in."

Try as they might, the mole babies were at odds again by Wednesday. The topic this time was whose whiskers were longest.

Mother Mole didn't even stop them. She simply held Fourth's hand and went for a stroll in the moonlight.

"Do you see the moon, Fourth?" she asked.

"It is haunting, Mother," Fourth replied. "It is not bright, and only a crescent, but it is very peaceful."

"Yes. Try to remember it exactly as it is right now. Now let's go back in."

It was the moon itself that caused the next squabble. It was Sunday again, and First declared that he had seen the moon exactly two weeks before and that the others had not been looking. "I saw it and it's round!" said First.

"No, it isn't, it's squashed on one side," Second shook his head.

"Wrong! It's only half of a circle," Third wiggled his whiskers confidently.

"I saw it and it was only a crescent!" Fourth whimpered while pacing back and forth.

"Mooooooother!" squealed Fifth. "I haven't even seen the moon yet. Please take me when you go."

Mother put her hand on Fifth's shoulder and took him out with her.

"Do you see the moon?" she asked.

"Not really," said Fifth. "Am I too blind to see it?"

"No," Mother said. "Nobody can see the moon tonight."

"It's gone!" cried Fifth. "All of our silly squabbling has caused it to run away. What have we done?"

Fifth ran back to the mole hole. "The moon is hiding. We made it go away with all of our arguing."

First through Fourth stared at him in disbelief. They ran out and looked frantically for the moon. It truly was gone.

Mother Mole found them desperately looking behind rocks and under bushes for the moon. "Don't worry about the moon," said Mother Mole. "We'll come out every night, and it will come back."

The first night they didn't see much, but after a few nights, they caught a glimpse of a thin crescent shape just before it set. Each night the crescent grew . . . and grew . . . until finally, after two weeks, it was a bright, peaceful, round ball again.

As they watched the full moon rising, Mother Mole asked them, "Now do you really SEE the moon?"

"Yes," they said. "The moon is sometimes bright . . ."

"And sometimes round . . ." said First.

"And sometimes squashed . . ." said Second.

"And sometimes half of a circle . . ." said Third.

"And don't forget that it can be a crescent . . ." said Fourth.

"And sometimes it is none of those things, " said Fifth.

"But it is always peaceful!" all five moles whispered, agreeing for once.

The Rock Cycle

Reading the Story

(large group or individual)

Materials:

- Copies of "Peter Morano" if reading independently
- Copies of "The Rock Cycle" worksheet, one per student

Either read "Peter Morano" aloud to the class or have students read the story silently. After reading the story, have the students draw pictures of Peter's journey. Make sure that students understand all of the words used in the story before they begin drawing. When they are finished, encourage students to share their pictures.

Mini-Lesson: Tectonic Plates

After explaining that Earth's surface is covered with plates of rock, give each student two flimsy paper plates cut into rectangular shapes so that they fit together better than circles. Talk about the fact that some plates move slowly apart. Ask kids to see what happens when plates move apart. What might happen if the plates slowly move apart? Talk about how the magma underneath the plates now has an opportunity to ooze up into the created space. Have your students ever seen a row of volcanoes? How might they have been formed?

Just as some plates move away, other plates must press into each other. If the students press their plates slowly together, what happens? Some plates overlap, pushing the bottom plate down. Others wrinkle, making mountains.

Discuss the forces that must act upon rock to change it, both by breaking down or forming new rock. Make a classroom chart of these forces. Encourage students to read about these forces and then share what they learned.

Inquiry Journal Activity

(small group)

Materials:

- "Inquiry Journal" form, one per student
- Rock sifters—These can be purchased at hobby shops, or use a colander (sifts out pebbles), a window screen (sifts out medium-sized particles) and the window screen with another square of screen on top of it (sifts out small particles). Silt-sized particles will be left.
- Measuring beakers or cups—It is easy to find the percentage of dirt sifted out at each stage if initial samples are approximately 100 mL.
- Geological map of your state

General Suggestions:

In your classroom, display the geological map and talk about some of the geological formations. Ask students who have visited these sites to describe them. After your discussion of tectonic plates, ask your students to explain how these formations might have been formed. Let the students use the discussion information when filling in the "Present Knowledge" section of the Inquiry Journal form. Their prediction about how dirt samples may vary should be based upon the ages of the formations in your area and their knowledge of how these formations came to be.

Follow the discussion about the formation of new rocks with questions about where dirt comes from.

Information: Erosion

Mechanical and chemical forces break apart rock. Mechanical weathering breaks rocks into smaller chunks without changing their mineral composition. Expanding and freezing rock will cause joints in rocks to widen and break. Frost action also breaks rocks apart. Water in a joint expands when it freezes causing a rock to break up. Rivers are a very important source of erosion, as are ocean waves. Plant roots can split rocks apart as they grow larger and larger. Tiny mosses and lichens loosen individual grains of rock.

Chemical weathering changes the minerals that make up the rock. Certain minerals combine with oxygen in the air or in the water to form oxides. Rain and groundwater can combine with carbon dioxide to form carbonic acid. Limestone is dissolved by carbonic acid and can form beautiful cave and cave formations.

In the "Design" section of the Inquiry Journal, consider how meaningful or interesting dirt samples can be obtained. Will students need to visit these places to get the samples? Can they write to other schools in your state and ask them to send samples? Will the class just go on a walking tour near your school? Make sure that the students plan how they will collect the dirt samples and document where the samples were found. Was the dirt sample collected near a river, a mountaintop, or in foothills? Is there any mining or farming nearby? Ask students to think about what information is significant when collecting data.

Your students may choose to look at the characteristics in the "Data Collections" table on page 42 or make up their own form by adding or deleting some characteristics. Remind them to carefully fill in their tables, so that they can compare the same traits with each sample.

To present their data, students can make a pie chart, showing the volume of each sample: pebble, medium-, small-, and silt-sized particles. If samples were 100 mL initially, the percentages of each type of dirt will simply be the volume. For instance, if a 100 mL sample has 15 mL of pebbles, that is 15% of the volume of the sample. You may find that samples with a large percentage of pebbles may shake out to parts that add up to more than 100 mL. This is because before the samples were sifted, the small particles filled in the gaps between the pebbles. If the sum of the parts is very different from 100 mL, you may have to figure the percentages based upon that sum of the separated piles. If the difference is minor, estimate the separated volumes down until they add up to 100.

When students tape their pie charts on the geological map at the sample source. Ask all groups to use the same color for each particle-size section on their pie charts. The charts will make immediate visual sense and patterns will be more obvious. When comparing, students should see that there are more large particles in dirt samples collected nearer to mountains or any "new" rock that is eroding. There should be more small particles in samples found near rivers. Ask the students to think about erosion while explaining this pattern in the "Final Analysis/Conclusion". They may also write down other ideas for further questions (investigations) they might explore. What would the students find if they collected dirt samples from other states? What would they find in other countries? What would they expect to see if they collected dirt from the top, middle, and bottom of a mountain? What does dirt on the bottom of a lake or the bottom of the ocean look like? What is the best kind of dirt for farming? Where might they look for fossils in your state?

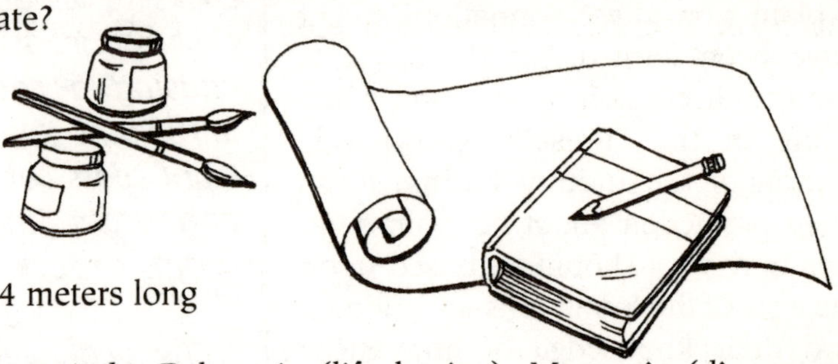

Time Line Mural

(group or individual)

Materials:

- Bulletin board paper approximately 5.4 meters long
- White paper
- Reference materials on geologic time periods: Paleozoic (life begins), Mesozoic (dinosaurs), Tertiary (giant mammals), and Quaternary (present)

For this mural, have each student choose a plant and an animal from a time period to research. Have the students draw or paint their animals and plants on the white paper. If working in groups, assign each group a time period to research. Depending on the number of students, you may want to assign two groups of four students to work on each time period. If so, make sure the students communicate with each other so that nobody is doing the same animal or plant. The dimensions for the mural are provided on page 43. It is recommended that the animals and plants for the present period are placed off the time line so that it is clear that the strip really is only 2 cm long.

Ask students to make two copies of their plant or animal, one to glue onto the mural and one to tape under the mural. That way, passing students can learn interesting information about the animal or plant while looking for that animal in the mural.

Rock Formations Drama

(small group or individual)

Materials:

- Rock samples that include igneous, metamorphic, and sedimentary rocks
- Construction paper, glue, scissors, markers, and tape
- A large chart with information as shown in diagram above

Rock Name	Igneous, Sedimentary, Metamorphic	How Rock Was Formed	Appearance	Uses

Give each group rocks to research. Ask each group of students to fill in the chart for each rock and present the information to the class. Now collect all the rocks. Give each group a rock and ask them to act out the formation of this rock. They may use four pieces of construction paper, glue, scissors, markers, and tape. Give the groups 10 minutes to plan a "dramatic representation." After a group performs, encourage the other students in the class to guess what rock the group had. After all groups have performed, give each group a new rock, repeating the steps until interest wanes.

Research Resources

Books

Barnes-Svarney, Patricia L. *Born of Heat and Pressure: Mountains and Metamorphic Rocks.* Hillside, N.J.: Enslow Publishers, 1991.
This book describes the formation of different types of mountains and metamorphic rock, and discusses their impact on their environment.

Barnes-Svarney, Patricia L. *Clocks in the Rocks: Learning About Earth's Past.* Hillside, N.J.: Enslow Publishers, 1990.
This book discusses the divisions of time used by geologists to record Earth's history and includes information on fossils, radioactive dating, and prehistoric life.

Cole, Joanna, *The Magic School Bus: Inside the Earth.* New York: Scholastic Inc., 1987.
On a special field trip in the magic school bus, Ms Frizzle's class learns about different kinds of rocks and the formation of the Earth.

Gallant, Roy A. *Our Restless Earth.* New York: F. Watts, 1986.
This book discusses the evidence that has led geologists to formulate certain theories about the formation of the Earth and the continual changes that have been taking place ever since.

Prager, Ellen J. *Sand.* Washington, DC: National Geographic Society, 2000.
This book describes the formation of sand from materials such as coral, rock, or crystals and shows how it can be moved through water, wind, ice, and other erosion agents.

World Wide Web Sites

http://www.usatoday.com/weather/wteach.htm
This links to other earth science teaching lesson plans as well as information about science standards.

http://www.proteacher.com/110073.shtml
This links to other earth science teaching sites.

http://www.ucmp.berkeley.edu/exhibit/geology.html
This is a fun place for students to get information about geologic time as well as providing links to teachers; provided by the University of California Museum of Paleontology.

Name ______________________________

The Rock Cycle

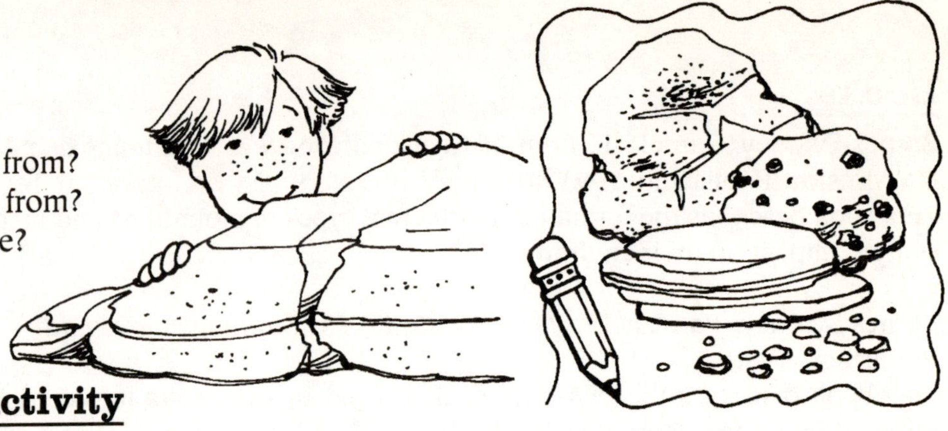

Think About It!

Where do rocks come from?
Where does dirt come from?
What does dirt become?

Inquiry Journal Activity

Best Question: Is all dirt the same?

Present Knowledge: Tell what you know about rocks. What forces change rocks? Looking at maps of your state, what formations do you see? How old are the rocks that make that formation? Where would the dirt there have come from?

Prediction: From the map and your discussion, make a prediction about what you expect to find when comparing dirt samples collected from nearby areas.

Design: When you plan your design, decide where you will go to get dirt samples. What characteristics will you compare? Particle size and visual characteristics are two traits that you may want to compare. You can separate your dirt sample into piles of similarly sized particles by sifting your dirt using screens with smaller and smaller holes. You can easily identify the color and texture of your dirt samples.

Data Collection: Your data collection table may look something like this chart.

Sample location and nearby geological formations	Volume of sample made up of pebbles	Volume of large-sized particles	Volume of sand-sized particles	Volume of silt-sized particles	Color	Texture

Name ________________________________

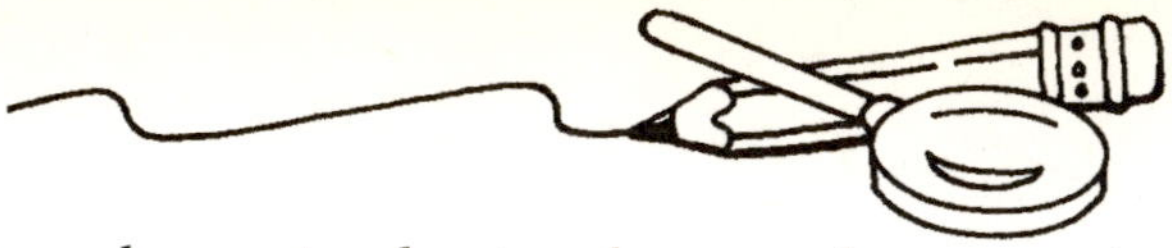

Data Presentation: When you have finished, make a pie chart or bar graph comparing the volume of pebbles to large particles to sand-sized particles to silt-sized particles. Tape your chart on the map showing where the sample was collected.

Final Analysis/Conclusion: When you have many pie charts taped on the map, compare them to see if you find similarities. Are samples near streams similar? Are samples near new rock formations similar? What might have caused this pattern? If you don't see any patterns, what might explain that?

Time Line Mural

When Peter entered geological time, things slowed down. Make a mural showing geological time where each centimeter of the mural's length represents one million years. If working in groups, choose an era to study. Each student should contribute at least one animal and one plant to the time line. Draw or paint pictures of your animal and plant. When finished, have a photocopy made of the drawings. On that copy, be sure to include the name of the plant or animal and a few facts about it.

The following information shows the four time periods with the approximate years that they cover and the amount of space needed on the mural:

- Paleozoic Era: 570 million years ago (mya)—245 mya (3.25 m long section)
- Mesozoic Era: 245 mya—70 mya (1.75 m long section)
- Tertiary Period: 70 mya—2 mya (68 cm long section)
- Quaternary Period: the last 2 million years (2 cm long section)

If you have chosen the Quaternary (present time period), how will you solve the problem of having only 2 centimeters of space on the time line?

Arrange with your teacher to display the bulletin board paper on a large open space. Glue the drawings of the animals and plants in the appropriate space. Display the second copy of the drawings beneath the mural for the viewer to read.

Rock Identification Drama

You will need to research the formation of several rock samples given to you. Fill out a rock chart, similar to page 42, with the information you gather. You will present what you have learned to your classmates, so everyone is familiar with all the rocks.

Now you will be given a secret rock. Your group must act out its formation while the class tries to guess which rock your group has received. You may use four pieces of construction paper, markers, scissors, and tape to help tell the story.

Peter Morano

Peter wasn't happy because he thought everybody bossed him around. He felt like his parents ordered him around at home, and his teacher ordered him around in the classroom. During recess, Billy the Bully was horrible to him. Sometimes bus drivers, waiters, coaches, and even strangers told him what to do. And on vacation, he could not get away from it all either. For example, his great-aunt Martha told him to tie his shoe just as he was climbing into the car for the drive home. Speeding down the highway, Peter's mother reminded him to tie the shoelace too.

He sighed and looked out the window in time to see a truck whining by.

"Hay truck, hay truck, bales of hay,
Take my wish and zoom away.
No one listens to me, Peter Morano,
I wish I were . . .
A big, bold volcano."

As the last words came out of his mouth, Peter became a volcano on the side of the road. Waiting until his parents' car was safely out of range, he gushed forth with oozing, molten lava, swallowing up the countryside around him. He rumbled. He smoked. He threw ash into the next state, and made sure that the roof of Great-Aunt Martha's house was blanketed in fine, powdered dust.

But, over time, Peter's interest in being a volcano subsided and the lava stopped flowing. No more hot gasses or volcanic bombs could be burped out. People walked all over his solid, black basalt and even carved their initials in it. Wind, sand, and water broke him down. Rain washed away his broken pieces in rivers to the sea. Peter's volcano wasn't powerful anymore. Sighing, he looked up in time to see a migrating bird flying freely over him.

"Winged bird with feathers gray,
Take my wish and soar away.
Being a volcano feels really poor
I wish I were . . .
A huge dinosaur."

As the last words of his wish came out of his mouth, Peter became a Tyrannosaurus Rex. He smiled with his big dagger teeth and roared. Swishing his tail through the air, he took off running across muddy beaches, leaving bird-like footprints behind him. Frightened animals ran, scattering right and left. He liked being big and strong. But he missed his old gang of friends. It would be nice to be able to play checkers with his mom before bed.

Many years passed, Peter was ecstatic to see another T. Rex moving into his territory. Peter ran to the new dinosaur, hoping for a friend.

"Hoooooooooooooowl!" The new dinosaur swished its tail at Peter, showing his fierce, dagger teeth.

"Oh, nooooooooooooo!" Peter howled back, and frowned with his older, cracked and blunted teeth. He recognized all the signs of a bully moving into his territory. Too old to win, Peter ran away, across the muddy beaches. Pushed out of the good hunting grounds, Peter the dinosaur felt very alone and wasn't powerful anymore. Many years after the dinosaur died, the rain washed the bones into the sea where they were buried by eroding volcanoes.

Knowing that wishes came in threes, Peter wanted to be extremely careful. He had already used up two wishes. It would take some time to come up with his last, perfect wish. Volcanoes loose their fire. Dinosaurs are ferocious but lonely. What did he really want?

As he waited and thought in his new peaceful surroundings, he changed. Other animals' bones washed over him, even the bones of the bully dinosaur. Minerals replaced the bones. Weight from ocean water, eroded volcanoes, fish skeletons, and clam shells squashed everything together. He slowly became limestone. It was so quiet and relaxing yet he still waited and thought.

After millions of years, the plate of Earth upon which he had formed rammed into another plate. It was pushed down toward the center of Earth. Intense heat and pressure changed him into marble. Marble. He thought about the beautiful marble statues he had seen in books and at art galleries.

It was time for his third wish.

"Swirling magma, hear what I say
Take my wish and ooze away.

I don't want to be a dino or volcano,

I wish I were . . . Peter Morano."

Even as he made the wish, he became himself again, sitting in the car with his parents. "Peter, did you tie that shoelace?" his mother asked.

"I'm tying it now," he said. After all, his shoe did need to be tied. He was glad to have a mother to care for him. When his dad reminded him to unpack his suitcase, he just said, "Okay." After all, didn't everybody have responsibilities?

At school, when Billy the Bully tried to tell him what to do, Peter just walked away, feeling sorry for the bully. He'd been as powerful as a volcano and a Tyrannosaurus Rex. He knew that Billy wasn't all that powerful after all.

From that day on, Peter became interested in rocks and sculpture. He read and practiced and read and practiced some more. When he grew up, Peter carved marble statues. His statues of dinosaurs and volcanoes and fish in the ocean made people pause and think about the beautiful Earth. His statues of loving friends and family made people smile.

The Water Cycle

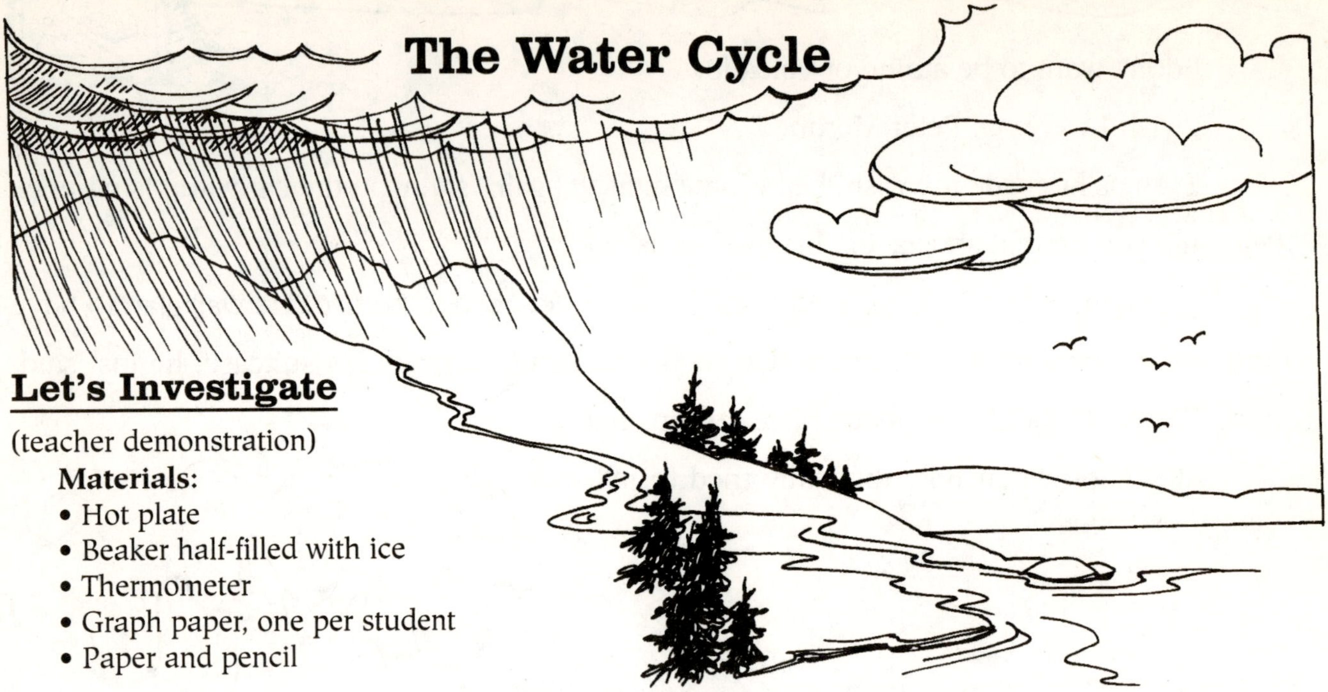

Let's Investigate

(teacher demonstration)

Materials:

- Hot plate
- Beaker half-filled with ice
- Thermometer
- Graph paper, one per student
- Paper and pencil

Ask students to make a prediction about how the temperature in the beaker will change during the next hour. Write their predictions on the board. Take an initial temperature reading. Place the beaker on the hot plate and turn it on high. On the board draw a table to record temperature readings. Ask different students to take temperature readings and record them on the table. Have students take Fahrenheit and Celsius readings for comparison.

Since there will be a great deal of waiting time during this demonstration, have the students generate a list of words related to water. Talk about the three states of matter (solid, liquid, and gas) and how water changes from one form to another. Be sure that all key words are included in your list of words. (Also talk about the water cycle: evaporation, condensation, and precipitation.) If there is extra time, encourage the students to use the words from the brainstorm list in a diamante poem (see page 49).

When you have finished collecting the temperature readings, ask students to draw a blue line over the part of the graph lines that represent the phase change from solid to liquid. What kind of line is it? Now ask them to draw a red line over the part of the graph line that represents the phase change from liquid to gas. Did the water get hotter during the change?

Reading the Story

(large group)

Materials:

- Copies of "Where Is the Rain?" if reading independently
- Copies of "Cycle Story Builder," one per student
- Drawing paper, markers, colored pencils, tape

Either read "Where is the Rain?" aloud to the class or have students read it silently. Talk about the water cycle (evaporation, condensation, precipitation, transpiration, and sublimation). Discuss with the students where these phase changes exist in the story "Where is the Rain?"

Now the students are ready to make a cycle story. This is a story that makes a circle and can be folded up, so any page can be the first page. The students, sitting in a circle, tell one part of the story. This story will be about a drip of water. The drip changes phases each time a student adds a new part.

Begin by having each student briefly tell his or her part of the story. Immediately afterwards, have the student jot down the main idea on the back of her "Cycle Story Builder" page along with the number indicating the order in which the story was first told. These notes will NOT be part of the book, but will help get the story into a coherent form. If students forget their part or order, it is very difficult to reconstruct. When the story has been told, ask the students to go back to their seats and write their part of the story on their "Cycle Story Builder" sheets, adding three details to make it interesting.

It is important that you have an even number of pages in order to make the story fold properly. It will be necessary to add an additional page if you have an uneven number of students.Tape the pages in a long line and then tape the ending page to the beginning page. Read the story the way it is written. As each student's part is read, that person must stand up and act it out.

Mini-Lesson: Poetry

In the case of phase changes, a good form for a poetry to express the concept is diamante, since it shows a change from one form to another. Here is an example of a diamante:

Ocean
mysterious, unknown
rolling, crashing, splashing
loudly, exuberantly, gracefully, silently
floating, gliding, flying
billowy, cottony
Cloud

As you can see, the poem starts by describing one thing and then changes halfway through to describe something new. The form is one noun, followed by two adjectives, followed by three verbs, followed by four adjectives (on this line the change occurs), followed by three verbs describing the new noun, followed by two adjectives describing the new noun, followed by the new noun.

Inquiry Journal Activity

(small group)

Materials:

- "Good Experiment Guidelines," one per group
- Inquiry Journal form, one per student
- Students will inform you of their needs. Let them be in charge of finding as many supplies as they can. If possible, find what they cannot. Some experiments may need to be changed if it is impossible to locate the equipment.

Students are asked to create an experiment that answers a question about phase changes. They may use the questions on page 53, or they may come up with their own questions. When they turn in their "Good Experiment Guidelines", read it carefully to make sure that they really will answer the question they asked. If the experiment the students have planned is impractical or will not answer the question, this would be a good time to talk to reassess their choice with them.

Boxed Water

(small group)

Materials:

- Cardboard or poster board
- Magazines to cut pictures for collage
- Tape to make boxes, glue to make collages
- Calculators (optional)

There are many choices to find a good scale to use while making the box to hold the Earth's water, but 1 cm = 10 miles will make boxes of a manageable size. Show them a units cube from a base 10 math manipulatives set if you have one, and explain that each one of those cubes stands for a cube of real water that is 10 miles long, 10 miles wide, and 10 miles high. Explain that you are going to be estimating and so they can round numbers. Have older students try to figure out how to make the boxes on their own. Hopefully, groups will come up with different solutions. For younger students, give them these directions to make cubes for each kind of water:

Type of Water	Dimensions of the Box
saltwater	68 cm x 68 cm x 68 cm
fresh water	12.5 cm x 12.5 cm x 12.5 cm
ice	19 cm x 19 cm x 19 cm
atmosphere	1.5 cm x 1.5 cm x 1.5 cm

Notice that liquid water has been separated into fresh water and saltwater. That is because the importance of fresh water is very different from the importance of saltwater. To get more information about how we use water, visit the USGS Web site: <http://ga.water.usgs.gov/edu/earthhowmuch.html> where these figures were found.

After decorating their boxes (or papers) students will present their boxes to the class and read the slips of paper reporting the importance of each phase of water. Afterwards, draw attention to the fact that there are many more slips of paper in the fresh water box but it is a VERY small box. How can we protect and conserve our fresh water?

Research Resources

Books

Cole, Joanna. *The Magic School Bus at the Waterworks.* New York: Scholastic Inc., 1986.
Ms. Frizzle takes her class on a field trip to the waterworks. They experience the water purification system from the inside of the operation.

Glaser, Omri. *Round the Garden.* New York: Harry N. Abrams, Inc. 1999.
Trace the journey of a tear as it falls to the ground, evaporates, reappears as rain, and waters a garden to make an onion grow to produce more tears.

Graham, Joan Bransfield. *Splish Splash: Poems.* New York: Ticknor & Fields Books for Young Readers, 1994.
This is a collection of poems celebrating water in its various forms, from ice cubes to the ocean.

Locker, Thomas. *Water Dance.* New York: Harcourt Brace, 1997.
Water is seen as storm clouds, mist, rainbows, and rivers. The book includes factual information on the water cycle.

Schmid, Eleonore. *The Water's Journey,* New York: North-South Books, 1990.
Explains the water cycle from precipitation, through brooks, to rivers, to an ocean, and then evaporation back into the atmosphere where it will repeat the water's journey.

World Wide Web Sites

http://www.broadwaters.fsnet.co.uk/index.htm
This site takes you on a virtual tour of a pond and shows how the water cycle affects pond life.

http://ga.water.usgs.gov/edu/earthhowmuch.html
One stop shopping, this USGS site has everything you will need to know about the water cycle. Explore the whole menu for a wealth of facts and figures. Especially interesting are the "Water Use" and "Activity" pages.

http://www.education-world.com/a_curr/curr216.shtml
An interesting approach to water-related environmental education. This site offers many articles as well as a curriculum action guide you can download.

Name ________________________________

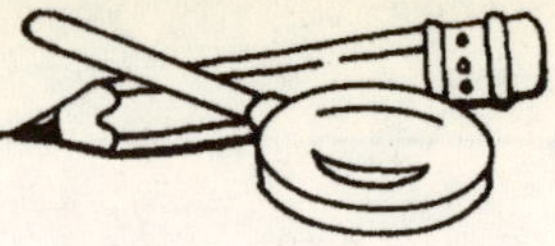

The Water Cycle

Think About It!

How much water is there on Earth?
Where is it found and in what phase?

Phase Change

While watching a demonstration about water phases, graph the temperature of the water as the phases change on a sheet of graph paper. What did you find out about the temperature of the water during the phase change from ice to liquid? From liquid to gas?

Reading the Story

Read "Where is the Rain?" Look for examples of the three phases of water: gas, liquid, and solid. Talk about the story with other students.

Now you are ready to make a cycle story, or a story that ends at the beginning. To make a cycle story, sit in a circle and decide who will start to tell the story. Your story will be about a drip of water. The person who starts the story can place the drip anywhere— in the air, in the ocean, in a plant, or in a snowman. Every person must include a phase change in his or her part of the story. When you finish telling your part, write it down on the back of your "Cycle Story Builder" sheet. Each student should write a number that tells where his piece of the story fits in with the rest. The last student must take the drip back to where it originally started.

Using the "Cycle Story Builder" sheet, write your part of the story in your best handwriting. Add three details to make your part of the story interesting. What did it feel like? What did it see? What did it hear? Finish with an illustration.

When the book is taped together in a circle, you can read the story many times, beginning on any page. As your teacher reads each page, the student who wrote that part of the story must stand up and act out the story line.

Each time you read the story start on a different page in the book. Delete a detail from each page with each reading until the story becomes very streamlined. See how fast you can tell the story and act it out.

Name ______________________________

Let's Investigate!

Each group will ask a question about phase changes and decide how to answer it by conducting an experiment. Follow the "Good Experiment Guidelines."

Questions to consider:

- Does water freeze more quickly or more slowly if you add other things (like salt) to it?
- Will water evaporate more slowly or more quickly if you add other things (like salt) to it?
- How can you make water condense on the side of a glass more quickly?
- Does the surface of the container affect how much or how quickly water condenses upon it?
- Do plants help water to evaporate or do they trap water, making it harder to evaporate?
- Which has more water in it, a cup of snow, a cup of ice, or a cup of water?
- If you melt a cup of snow, will there be more water in snow gathered from different sites?

Fill out the Inquiry Journal pages included in your packet as you work.

Boxed Water Models

How much water is there on Earth? Using the information below, show approximately how much water is in each of the three phases by making a box to hold each one.

Gas: (water in the atmosphere)—3,100 miles^3

Liquid: 1) saltwater (oceans and inland seas)—317,025,000 miles^3
2) fresh water (rivers, lakes, and groundwater)—2,046,300 miles^3

Solid: (polar caps and glaciers)—7,000 miles^3

To make a box that will fit in your classroom, use the scale 1 cm = 10 miles. Remember, volume = length x width x height. On strips of paper write why this phase of water is important to us. If you cannot explain the reason, do some research. On the side of each box, create a collage using pictures from magazines, paints, and colored sheets of construction paper. Include the strips on which you have written about the importance of this phase of water. If the box is not big enough to decorate, make your collage on a piece of construction paper and glue your box to the collage.

Name ______________________________

Good Experiment Guidelines

What question are you asking? ______________________________

____

In order to have an experiment that tests your question, you must make sure that only ONE thing is being tested during the experiment. What is the ONE thing that you are testing in this experiment? How are you changing that one thing?

____

____

____

List all the things that are staying the same during this experiment. Use the words, "natural conditions," to include everything like gravity, the speed of Earth spinning through space, etc. Will your containers all be the same size? Is the water from the same source? Are all steps of the experiment being done on the same day so that humidity or air temperature does not change?

____

____

Do you need a control, in other words, a sample that is not being changed in any way to compare with the sample that has been changed? How will you prepare it?

____

____

Request for Supplies

Items I Need	How Many	Purpose

Name ______________________________

Cycle Story Builder

Detail 1: ______________________________

Detail 2: ______________________________

Detail 3: ______________________________

Where Is the Rain?

At first, when rain didn't come, Iris and her family bought water in jugs. But soon they had spent all of their money so they buried the last jugs just like a treasure. Their orchard of peach trees seemed to stand guard with their gray-brown leaves curled and dying.

Iris and her family were careful. They only dug up a jug when they were desperate. Even so, now there was only one jug left.

One day, as Iris slept in the orchard, she dreamed the trees were covered with rich green leaves and fuzzy peaches. Taking one, she bit into it. The sugary juice streamed into her mouth and down her chin. Suddenly she awoke, surrounded by a dust devil that threw sand everywhere. The whirling wind seemed to whispered, "Someone is needed to go on a journey to bring back the rain. Who will go?"

"I'll go!" Iris shouted and ran into the swirling sands.

"Will you take me to the sky?" she asked the dust devil. A gust of wind lifted her into the air, then dropped her on a soft white cloud.

"Sky, please help me," Iris called. "Your breath holds the water that we need. Please give us some of your rain."

"Child, child," rumbled the sky. "If only it were so simple. Breathe deeply and feel my dryness in your lungs. If you want rain, you'll have to go and ask the ocean to give me water. I cannot give what I do not have."

Iris was swirled around and down and tossed back onto her parched land. "The ocean! The ocean is too far away. I'll never make it without anything to drink." She ran to dig up the last jug of water. "One jug isn't enough to quench my family's thirst," she said, "But it can help me bring back the rain." She clutched the jug and began to walk. On her journey she saw shrunken and thirsty lizards and many completely dried-up bushes and trees. She saw parched people who needed her jug of water. Finally, the breeze brought the cawing of sea gulls and the scent of seaweed.

"Ocean! Please help me! Will you evaporate some of your plentiful waters to the sky so that it can give us the rain we need? Please?"

"Child, child, if only it were so simple." The ocean's waves crashed. "I must keep a delicate balance of water and salt to nurture the kelp, coral, fish, and whales. If I gave the sky my water, I would become a salty stew of death. You must go and ask the river to give me fresh water. I cannot give what I cannot spare."

"But all the rivers are dry," Iris protested.

The ocean's waves only crashed in return.

"The rivers are dry," she repeated, "like my jug, only a trickle left." Iris turned away, trudging to he river.

At the river's banks, Iris heard no happy gurgling, just the sharp buzzing of flies. She smelled decaying fish.

"River! Please help me!" she begged. "Tell me what I can do. We need rain. But the ocean must have fresh water from you in order to give water to the sky, so that it can give us the rain we need. But you have no water."

The river sighed. "Things are never simple. You must go and ask the mountain to give water to me. The mountain's snow-capped peak has the water we need."

Iris dragged the jug to her cracked lips. She swished the last of the water in her mouth to make it seem like more. It would be a hard climb.

Setting out on wobbly legs, she steadied herself by holding onto trees. She leaned against their trunks, careful not to sit. Near the mountain's peak, she eagerly ate some snow. It only made her more thirsty. When she could go no farther, she lay in the snow.

"Mountain, please help me," she said softly. "Will you let us have some of your snow? The river must have water to give to the ocean, so that the ocean will give water to the sky, so it can give us the rain we need."

"Child, child, it's not that simple. I like my snowy white coat. I have decided to keep it. The river only wastes my waters anyway. Greedy plants and animals take it all and give me nothing in return."

"I think you're wrong, " Iris pleaded. "Plants and animals do give water back. The water given back mixes with water evaporating from the river and the ocean. The sky uses that water to make fresh crystal snowflakes for you each winter. In the spring, you must give back your coat, or you will never have another new one."

"I do not trust you, girl," the mountain replied. "You ask me to give, give, give. Yet you are not willing to give at all."

"This is all I have," Iris said, tipping the jug upside down. "It is all the water from my home." One drop, her last drop of water, fell onto the snow. Its warmth melted and a small tunnel was made down to the hard rock below. The mountain's heart was touched.

"Let the sun begin its job here," The mountain said. Light poured into the tiny tunnel and began melting the mountain's snow. Trickles of water ran together forming brooks which emptied into the river. The river swelled, sweeping up Iris and running to the ocean. The wind sent Iris up to the sky. The sky formed a black rain cloud that took Iris back to her home. She was set on the ground as the rain began to fall. She was so excited as she ran to her parents. As they hugged, huge drops of rain plopped in the dirt.

The Carbon Cycle

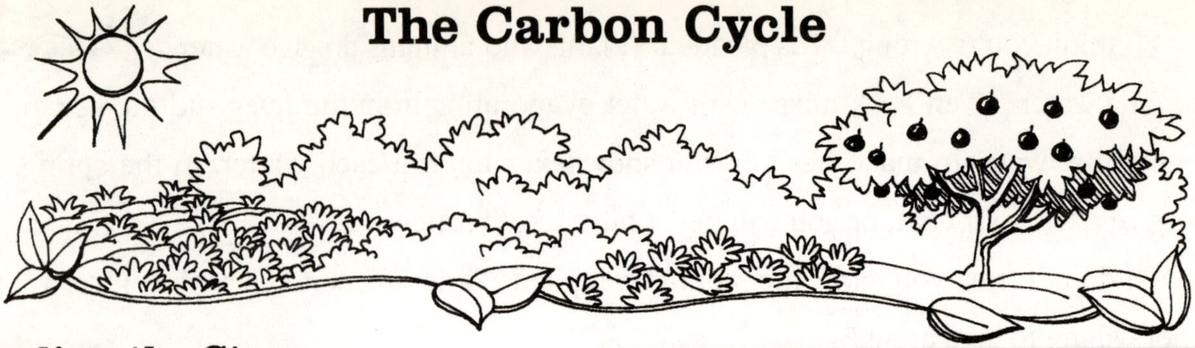

Reading the Story

(large group and/or small group)

Materials:

- Copies of "Eating Sunlight" if reading independently

Either read "Eating Sunlight" aloud or have students read silently. Discuss the storyline.

Mini-Lesson: Tessellation

Art projects always work better if students can see what they are trying to achieve. Check out some books from the library with M. C. Escher's tessellated drawings and paintings. Show the students how the negative space makes a shape of its own in the work. In the simple tessellation shown on this worksheet, the negative space does not create a recognizable form as it does in Escher's work, but it is pleasing to the eye. If the students make shapes that almost fill up the hexagons, they may notice that some parts connect to each other as the grasshoppers' antennae connect to the feet. The blending of the shapes is pleasing to the eye and continues the metaphor of connections.

Story Activity

Provide copies of the "Backyard Food Chain" worksheet and the "Tessellation Food Chain" worksheet.

Complete the "Backyard Food Chain" worksheet. Relate the terms on the worksheet with the characters in the story. In the picture, the producers are the plants, the primary consumers feed on plants, and the secondary consumers feed on the plant eaters. The decomposers are difficult to show and would be represented by decaying plant matter.

Have each student choose an animal or plant from the story. Their choices are: Owl, Snake, Toad, Grasshopper, or plants Grasshopper ate. Have students complete the "Tessellation Food Chain" sheet. Have students connect their chain pieces in the order the animals or plants appeared in the story. See student page 64 for illustration.

Inquiry Journal Activity

(small group or individual)

Students will be working with mass, not weight of a tree, the difference being that mass is not influenced by gravity. It may be worth a few minutes to discuss the difference.

Materials:

- Brown paper for trunk and branches
- Yellow, red, orange, or brown paper if making leaves
- Glue and scissors
- Background paper, 9" x 12" (23 cm x 30 cm) for individual trees, or butcher paper for large trees
- Inquiry Journal form, one per student

General Suggestions

- **Best Question**: "What percentage of a tree is made from carbon dioxide in the air?"
- **Present Knowledge**: What do the students know about the composition of air? Do they know where carbon dioxide comes from? What do they know about photosynthesis?
- **Prediction**: How much of a tree is made from air only. Draw a sketch of the prediction, remembering to take into account the all aspects of the tree's life.
- **Design**: Once they find the percentage, how will they make a tree that illustrates it. They should discuss how they will go from percent to fraction and how they will measure that fraction part. How will they know how much of the tree to remove?
- **Data Collection**: This is the section where they should write down all of their calculations for determining the percentage and fractional part of the tree made from carbon dioxide.
- **Data Presentation**: Their tree is the presentation. If the tree is too large to put on the Inquiry Journal, they can direct the reader to the wall in the room where it hangs. Another possibility is to take a photo of the tree and put that in this space. There should also be a brief narrative, a long caption describing the meaning of the tree.
- **Final Analysis/Conclusion**: Were they surprised to learn how much of a tree comes from air? How might air pollution affect plants? When we burn fossil fuels, wood and coal we put carbon dioxide into the air? Could that be bad? What is the greenhouse effect and what does it have to do with the carbon cycle?

They will also calculate the percentage of a tree made from carbon dioxide in the air.

Step 2 is to determine how much of that sugar came from carbon dioxide and how much came from water. Start by discussing the studies done in 1648 by Nicolai de Cusa who planted a 5 lb. willow tree in an earthen vessel with 200 lb. of dry soil. He added nothing but water to the vessel. After 5 years, the willow tree weighed 169 lbs. The dried soil weighed 200 lbs. minus 2 ounces. He concluded that the growth, (164 pounds) was from the water. However, we now know the equation for photosynthesis: $12H_2O + 60_2 + \rightarrow 60_2 + C_6H_{12}0_6 + 6H_2O$ and we can see that 6 molecules of water were given back to the air. Where did the other six go? They donated their hydrogens to the sugar molecule. Their oxygens were used to form O_2, oxygen molecules that we need to survive.

Most of the dry weight of a plant comes from sugars (and compounds made from sugars) manufactured by the plant during photosynthesis. Now we also know that the carbons and oxygens were donated by the carbon dioxide and the hydrogens were donated by water. How can we use that information to calculate the percentage of mass that comes from air?

Since we know that there are small amounts of other elements like phosphorus and nitrogen in the proteins the tree has made, we will (generously) round this number down to 90%. This means that only 10% did not come from carbon dioxide. We could divide the tree into hundredths and leave 90 while cutting away 10. This however, would be very time consuming. Luckily, we can turn percentages into fractions. Therefore, to make our tree, we only need to cut away one tenth of the tree's trunk, one tenth of each branch and one tenth of each leaf.

Before letting students begin making their models of trees, make sure that their data collection includes these three main ideas (if developmentally appropriate):

1) the equation for photosynthesis and a brief explanation
2) the atomic mass units for the simple sugar made by the plant during photosynthesis and how much of the mass was donated by carbon dioxide and how much was donated by water.
3) the percentage of the mass of the tree that was donated by carbon dioxide and the percentage donated by water; include the fractional equivalent.

Mini-lesson: Elements, Atoms and the Periodic Chart

Hand out a copy of the periodic chart or display it on a wall. Give students these words: atom, molecule, nucleus, proton, neutron, and electron. At the end of the lesson, they will have to define the words. Begin by drawing a model of the atom with protons and neutrons in the nucleus and electrons orbiting. Establish that we call the mass of a proton one atomic mass unit (amu). Neutrons also have a mass of one amu. The mass of an electron is so relatively small that it is considered to be zero. The important thing to remember about electrons is that the number of electrons equals the number of protons. Now ask your students to find carbon, hydrogen, and oxygen. How many protons and electrons do these elements have? How much mass do they have? Draw a water molecule on the board and ask them to find out how much mass it has. They should add 16 amu from oxygen and one from each hydrogen to get a total of 18. Now find the atomic mass of carbon dioxide. One carbon has 12 amu and 2 oxygens have 32 amu for a grand total of 44 amu. Now find the mass of the simple sugar produced in photosynthesis, $C_6H_{12}O_6$. They should discover:

$$6(12 \text{ amu}) + 12(1 \text{ amu}) + 6(16 \text{ amu}) = 180 \text{ amu}$$

This sugar is the basic building block that makes the tree you see.

Make sure that students understand that the tree they made will be dead and dried out. Therefore, students can disregard the weight of water in the tree which would be considerable. This water is in the form of sap and carries nutrients around the tree in the same way that blood carries nutrients within our bodies. Decide whether you want the students to make small, individual trees or large, group trees. Start with rectangles that are 127 mm x 50 mm (5 in. x 2 in.) for the trunk, 102 mm x 13 mm (4 in. x 1/2 in.) for the branches, 50 mm x 6 mm (2 in. x 1/4 in.) for the smaller branches. For one large tree measure in meters. Small, individual trees should be measured in centimeters.

If you choose to make one large tree for the classroom, assign groups of students to be responsible for making different parts of the tree. One group of students can make and decorate the trunk, another group can make large branches, a third group can make smaller branches, and the fourth group can make twigs. All students can create leaves.

Extension: Dead trees contribute a great deal to the forest habitat. Visit the Web site listed in this section to learn about snags or read about them in books. Decorate your tree by adding plants and animals that will be using this dead tree for food or shelter.

Ecosystem Research Paper

(small group)

Since there are four topics of research to this paper, students should be placed in groups of four. Students will need to know how to fill out index cards, organize research, and make bibliographies.

Research Resources

Books

Asimov, Isaac. *How Did We Find Out About Photosynthesis?* New York: Walker, 1989.
Traces the scientific discoveries that led to our understanding of photosynthesis and how this process relates to the food supply, changing ecological balance, and threats to the Earth's atmosphere.

Ross, Bill, Ph. D. *Straight From the Bear's Mouth: The Story of Photosynthesis.* New York: Atheneum, 1992.
Mr. Mildew, an eccentric scientist, helps Dana and Jake set up a science project on photosynthesis.

Silverstein, Alvin. *Photosynthesis.* Brookfield, Conn.: Twenty-First Century Books, 1998.
The book explains photosynthesis, the process responsible for providing the material and energy for all living things, and discusses such related issues as respiration, the carbon cycle, acid rain, and the greenhouse effect.

World Wide Web Sites

http://library.thinkquest.org/11226/
This site follows Captain Carbon through the photosynthesis where he becomes part of the protein cycle or the carbohydrate/fuel cycle and back to carbon dioxide where he will go through photosynthesis again.

http://www.metrokc.gov/dnr/kidsweb/snag_stuff/snags.cfm
A wonderful site showing the animals that use dead trees to survive. This is a great site to use after making the 90% tree.

Name ________________________________

The Carbon Cycle

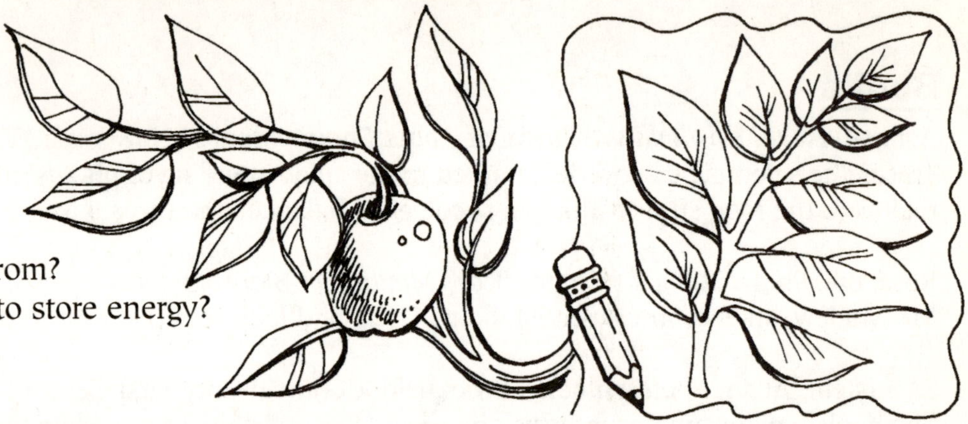

Think About It!

What is carbon?
Where does it come from?
How do plants use it to store energy?

Story Activity

After reading "Eating Sunlight," fill out the "Backyard Food Chain" worksheet. Complete the "Tessellation Food Chain" worksheet. Tape your strips to form a paper cycle chain.

Inquiry Journal Activity

Plants must have three important things in order to grow: sunlight, carbon dioxide and water. Imagine doing an experiment like the one done by Jan Baptista van Helmont who in the 1600s wanted to understand tree growth. He found that in five years, a tree added 165 pounds of weight but only used 2 ounces of dirt. From where did the other 163 pounds come?

Best Question: "What would a tree look like if we could see how much of it came from air?"

Prediction: Sketch how much of a tree is made only of air.

Present Knowledge Section: In this section think about how carbon dioxide gets into the air. What does the plant produce during photosynthesis? What else do you already know about photosynthesis.

Design: Percentages and fractions play an important role in the completion of this inquiry. In this section, discuss the meaning of percentages and how they relate to fractions. Once you have determined the percentage of the tree that is made from air, how will you find the fractional part? How much will you cut off? How much will you leave behind? What strategy will you use to remove that fractional part of the tree trunk, branches, twigs, and leaves? Also address the fact that the tree you make must be dead and dried up. This allows you to ignore the weight of water in the tree. Leaves must be brown and scattered around the bottom of the tree to indicate that they are dried.

Data Collection: Take notes while your teacher explains how carbon, oxygen, and hydrogen form simple sugars that store energy. Include all of the calculations you do to determine the percentage of the tree that comes from carbon dioxide. Do not forget to turn your percentage into a fraction and reduce it. Don't worry, your teacher will be helping you.

Name ____________________________________

Data Presentation: Your data presentation will be the completed tree and a written description of it. If your tree is too large to fit on the Inquiry Journal, refer the reader to see the tree where it is in the classroom.

Final Analysis/Conclusion: Discuss how closely your predicted tree matched your final tree. Were you surprised to see that so much of tree is made from air? Many people believe that we are putting too much carbon dioxide into the air. Why is that problem? What is the greenhouse effect? How can we stop putting too much carbon dioxide into the air?

Ecosystem Research Paper

Once the carbon dioxide in the air has been made into a plant, that plant will fuel the food chain of an ecosystem. Choose an ecosystem from the list below to research.

1) Open ocean
2) Continental shelf waters
3) Estuaries
4) Wetlands (marshes and swamps)
5) Lakes and ponds
6) Rivers and streams
7) Tundra
8) Boreal coniferous forests
9) Temperate deciduous forests
10) Temperate grassland
12) Tropical grassland and savanna
12) Desert
13) Tropical rain forest

Write a paper about the ecosystem by having each member of your group research one of the following topics:

- The climate (Those who are studying the ocean will want to discuss the temperature and light changes that occur giving rise to different life zones.)
- How plants have adapted to this ecosystem
- How animals have adapted to this ecosystem
- How humans (native peoples) have adapted to live in this ecosystem or, in the case of oceans, how people benefit from the life there

Take notes on index cards while doing your research. Restate the ideas in your own words before writing them down on a card. You may copy a sentence word for word if you intend to use it as a direct quote, but remember that you must give credit to the author with a footnote or endnote. Also, keep a bibliography card for each book or Internet site from which you take information so that you can make a bibliography of your group's combined resources.

Food Chain Activity

Along with your research paper, make a poster or diorama of a food chain that might exist in your ecosystem. Put the sun at the center of your food chain since it provides the energy that runs the food chain cycle. Don't forget bacteria and other decomposers connect all of the links in your chain to the plants that use their nutrients to make food. Be sure to show decomposers as a link in this cycle.

Name ______________________________

Backyard Food Chain

Define the following words and then label the picture below using the words:

Producers:

Primary Consumers:

Secondary Consumers:

Decomposer:

Name ______________________________

Tessellation Food Chain Links

1 Choose an animal or plant from the story "Eating Sunlight". Your choices are: Owl, Snake, Toad, Grasshopper, or plants for Grasshopper to eat.

2. Make the shape of your plant or animal in this practice hexagon. Try to make it fill the shape as much as possible:

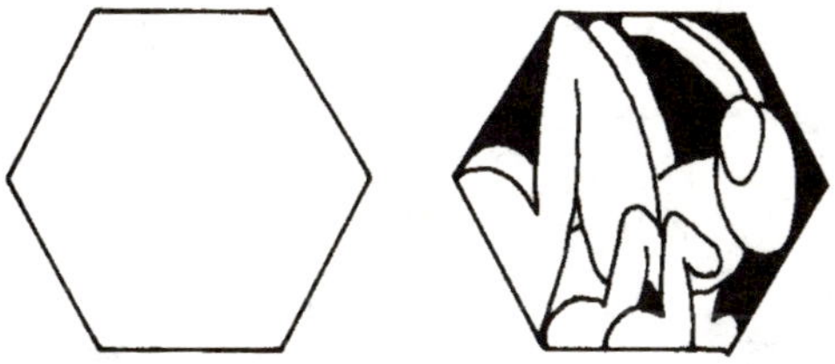

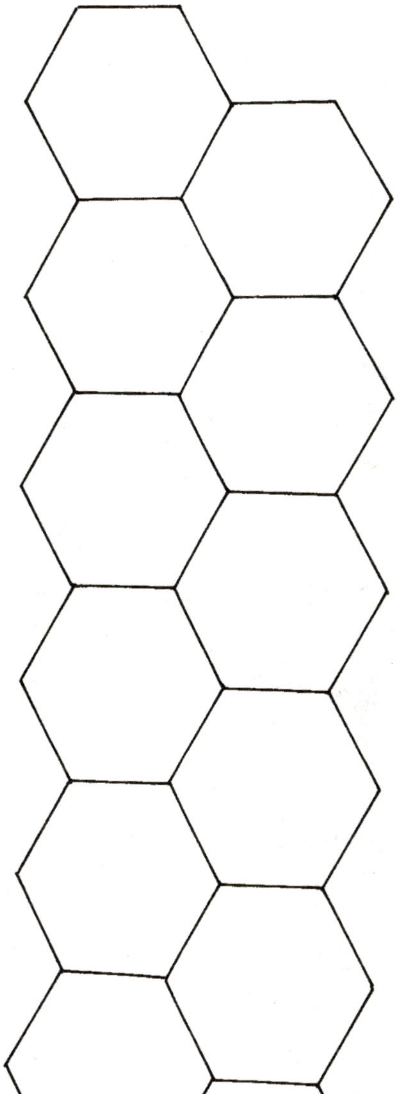

3. Cut out the animal or plant you have drawn in the practice hexagon.

4. Trace around its shape in each of the hexagons on the strip.

5. If you notice a place where the figures can touch, connect them. Don't worry if they don't connect, it will look great.

6. Color your animal or plant the same figure in each hexagon on the strip.

7. You may either color the background in each hexagon the same color or color alternate diagonal pairs the same color.

8. Cut out your tessellation food chain. Make a paper chain by connecting your strip with your classmates.

Remember: Be sure your chain pieces are put together in the order the animals or plants appeared in the story.

Eating Sunlight

"I feel like I'm getting old," Grandmother said. Her voice was soft and gentle.

"Don't worry," her grandson Phillip said. "I can take care of you."

Grandmother smiled. "You are growing up but . . . "

"I can do anything you want!" he declared.

Lifting her eyebrow, she looked at Phillip. "If that's so, then go and gather some sunlight for me. I am craving the taste of it."

Phillip looked at his feet. "Nobody eats sunlight! Nobody can gather it." Nevertheless, Grandmother had asked him to do it, and Phillip didn't want to admit defeat. Running outside he promised, "You can depend on me."

Out in the garden, the sun wrapped around him like a warm blanket. He tried to catch some sunlight in his cupped hands. However, when he peeked under his thumb, he saw just darkness. maybe he could catch sunlight with a bucket! But unfortunately, all he caught was darkness. He thought he might be able to catch some in a sack. Once again though, it was only darkness. Perhaps a hole would do the trick. Alas, there was simply darkness again.

Sad and sweating, he sat in the shade of the apple tree. He wanted to pick one crunchy piece of fruit, but there was no time. The sun, after all, was already going down. Phillip began to worry because there was no hope of catching sunlight after the sun went down.

Up in the tree, a branch swayed as an owl changed sleeping positions. Phillip jumped up. Owls are wise. Maybe Owl would tell him how to gather sunlight.

"Hello!" Phillip called.

"Whoooooo are you?" Owl blinked.

"I am hoping you can help me," Phillip began. "My grandmother says that I can gather sunlight. Is she right?"

Owl ruffled her feathers. "Your grandmother is correct. You came to the right place. I'm an expert at catching sunlight. Every night I swoop down on it, and snatch it, wriggling and writhing in my talons."

Phillip looked at his human arms. He knew he'd never swoop or snatch like Owl, but he wouldn't give up either. Then he thought about what Owl had said.

"Wait a minute, how can you catch sunlight at night?" Phillip asked. "It's dark!"

"That's true," Owl replied. "It is Snake that carries sunlight into the night for me."

Phillip said, "If that's so, then Snake must know how to gather it."

It didn't take long to find Snake weaving his way around the corn stalks. Phillip's mouth watered looking at golden ears of ripe corn. He didn't dare waste time picking any. The sun, after all, was going down.

"Hello, Snake, I've come to ask for help," Phillip said. "Owl told me you carry the sunlight into the night for her. "

"That'ssss right," Snake hissed. "I am a ssskilled hunter of sunlight. I sssneak up on it in the shade by the river, unhinge my jaw and ssswallow it whole."

Phillip frowned. He wasn't sneaky like Snake, and he'd never be able to unhinge his jaw. Besides, it just didn't make sense to him.

"Nobody can catch sunlight in the shade," Phillip challenged, hoping for more information. "It's too dark."

"Sssilly! Toad brings me sssunlight."

"Then I'll go see Toad," Phillip whispered.

He found Toad dozing in the shade of the tomato vine. Phillip resisted the temptation to eat one of the juicy, sweet tomatoes because if he didn't hurry, Grandmother would have no sunlight on her table tonight. The sun, after all, was going down. In the night, there would be no hope of catching any light.

"Hello, Toad. Is it true what snake says about you?" Phillip asked. "Snake says that you can catch sunlight faster than anybody."

Toad's wide mouth smiled. "Yep. My tongue brrrings it in."

"If only I had a clever tongue like you." Phillip shook his head.

"No despairrring. Grasshopper might teach you. He brrrings my sunlight to me."

"Oh, I don't know!" Phillip cried. "I'm not even sure Grasshopper can help me now. It is getting late and the sun is going down."

"Ask me," Grasshopper hummed from a deep-green spinach leaf.

Phillip slumped to the ground and watched the sun slipping away. "Why bother? My Grandmother is craving the taste of sunlight and I won't be able to bring her any. I don't have Owl's pointed talons. I don't have Snake's unhinging jaw. I don't have Toad's clever tongue."

"Still, you and I have what they need. We have our breath."

Phillip held his breath, then blew out fast. "My breath. How can I catch sunlight with my breath?"

"YOU can't," Grasshopper smiled.

"But you just said . . ."

"You see, breath is needed to catch sunlight. But YOU can't do the catching. Only plants can capture sunlight. They use it to make their leaves, stems, and seeds. But they can't do it without your breath."

"The plants use my breath to catch the sunlight? Then I gather the plants?"

"That's right. Plants are full of the sun's energy. Pick some spinach here in the garden and then repay the plant with your breath. That is what your grandmother wants."

Phillip picked some spinach, then thanked Grasshopper and headed for home. He stopped near Toad and picked two tomatoes. "Did you find what you were looking forrr?" said Toad.

"The sunlight is in the plants!" Phillip bubbled with excitement. "Does Grasshopper bring you spinach and tomatoes to eat?"

"No, Grasshopperrr eats plants. I eat Grrrasshopper. The sunlight is passed on to me."

"Then Snake gets the sun's energy from eating you. Owl gets the sun's energy from eating Snake. Wait until I tell Grandmother." Phillip raced to the cornfield. He picked two ears of corn and told Snake all about what he had learned.

Back at the tree, Phillip climbed up, picked two crunchy apples, and explained to Owl about the sunlight. "But Owl, there is still one thing I don't understand. You and I are not eaten by anybody. What happens to the sunlight in our bodies?"

"When we die, decomposers use the sunlight in our bodies and the bodies of other dead plants and animals to. They return our nutrients to the earth where plants use them to make new plants."

"It's a circle!" Phillip said. "Energy comes from the sun. Then plants catch the energy and pass it to animals who use it and pass it back to plants and so on . . ."

"True," replied Owl.

With his arms full, Phillip ran to the house. "Look, Grandmother!" he called. "I brought enough sunlight for both of us. On my journey I learned that I'm craving the taste of it, too. But you already knew that, didn't you?"

Recycling

Reading the Story

(large group and small group)

Materials:

- Copies of "Looking Backward, Thinking Forward" if reading independently

Either read the story to the class or have them read silently. You may want to encourage students to participate by reading different speaking parts. Have the students plan a campaign for the mayor of Garbageville. The main issue of the campaign is the problem of having too much garbage.

Story Activity

Have the students gather in small groups and decide who their candidate for mayor of Garbageville will be. Direct them to explore the Internet or read books about garbage and recycling before they write the speech for their candidate. Encourage them to think of new and creative solutions for solving the garbage problem.

If possible provide a video camera and tape for the campaign. When students have finished taping their mayoral speeches, give the video to other classes to view. Ask them to vote for mayor of Garbageville. Leave a space on the ballot for voters to explain how they made their decision.

Inquiry Journal Activity

(small group and individual)

Materials:

- Inquiry Journal form, one per student
- Graph paper for data presentation for bar graph or white paper for pie chart

General Suggestions:

In filling out this inquiry, it is important to have a discussion about honesty. Students may want to recycle waste in their lunches or meals so that they have smaller amounts of trash. Be sure that students keep track of all categories of trash: paper, plastic, metal, and glass. If some students do not presort their trash at home, encourage them to begin immediately.

What measure will they use? If students choose to use cubic centimeters, ask them to make a 10 cm x 10 cm x 10 cm box (1000 cm^3) to help them estimate small amounts like lunch garbage. Would their lunch garbage fill up one box (1000 cm^3)? Would it only fill up one-half box or 500 cm^3 ? Two boxes (2000 cm^3)?

When students have filled in the table with data on their garbage production during one week, ask them to add all categories together to find out the total volume. Then, find the percent of garbage in each category. If your students are not ready to understand percents, you may want to make a simple bar graph instead.

In order to answer the question completely, ask each student to multiply the amount of garbage going to the landfill by 52 weeks in a year. The total will be too large for students to understand the size of the quantity. Measure your classroom to find its volume. Divide the yearly total by the classroom volume to find out what part of the space in the classroom could be filled by one student's garbage. Ask the students to discuss this information and what it means to a town the size of your town in the "Final Analysis/Conclusion" section of their inquiry.

Mini-Lesson: Mean, Median and Mode

Find out the total volume of garbage that each student is sending to the landfill. Ask them to put the number on a slip of paper anonymously. Put all the numbers up on the board. Use these numbers to find: mean—the average, median—the number halfway between the highest and lowest, and mode—the number that comes up most often.

Mini-Lesson: Volume

Review the formulas used to determine volume. Besides the volume of rectangular solids, students may now choose to estimate garbage sacks as right cylinders. Talk about the concept that all volumes are found by first determining the base of the object and then multiplying that by the height.

Rectangular solids then become:

(length x width) x height = volume

(area) x height = volume

How will they use this information to measure lumpy, bumpy garbage?

Mini-Lesson: Percent Review

If students are measuring their garbage in cubic centimeters, you can use a base-10 hundreds flat to illustrate percents. Unit cubes represent one cubic centimeter of garbage. If you could compress all of your garbage to the size of one hundreds flat worth of garbage then we could compare the number of unit cubes of each kind of garbage. The way we do this mathematically is by finding percents.

How many cubes would represent plastic garbage? How many would represent paper? How many would represent steel? How many would represent glass? How many would represent garbage going to the landfill? Pie charts take this information and put it in the shape of a circle instead of a square.

Research Resources

Books

Condon, Judith. *Recycling Glass*. New York: Franklin Watts, 1991.
This book explains how glass is recycled.

Condon, Judith. *Recycling Paper.* New York: Franklin Watts, 1990.
Paper recycling is discussed.

Palmer, Joy. *Recycling Metal.* New York: Franklin Watts, 1990.
Metal recycling is explained.

Palmer, Joy. *Recycling Plastic*. New York: Franklin Watts, 1990.
This book explains how plastic is recycled.

50 Simple Things Kids Can Do to Recycle. The Earth•Works Group. Berkeley, CA: Earth Works Press, 1994.
Ideas are given for recycling projects at home, at school, or anywhere.

World Wide Web Sites

http://www.reversegarbage.org.au/re-usables.htm
Reverse Garbage is a company based in Sydney, Australia that promotes reusing instead of recycling by providing clean garbage and education about how you can reuse what might have been thrown away.

http://www.metrokc.gov/dnr/kidsweb/solid_waste_main.htm
Incredibly informative and kid-friendly site this is about how much garbage we make and how we can make less.

http://members.aol.com/donnandlee/Green.html#RRR
You can surf for hours at this site set up to help teachers teach about Earth Day. It offers wonderful recycling activities.

http://www.ciwmb.ca.gov/popups/energy1.htm
This is a page set up for California's waste management. It is not posted for student use, but it does show what kinds of programs are in place or envisioned to manage the waste problem.

Name ______________________________

Recycling

Think About It!

- What kinds of trash do you recycle?
- What kinds of trash are reused?
- How can we recycle more of our trash?
- How can our recycling efforts mimic the closed cycles we find in nature?

Inquiry Journal Activity

Best Question: How much of your garbage do you recycle in a year? Getting an exact figure isn't possible, but in this inquiry you can make a rough estimate.

Present Knowledge: Think of all the kinds of garbage that you make. Where do you throw your garbage away? In the cafeteria? In your classroom? At restaurants? Do you throw away more garbage at certain times of the year? What kinds of garbage do you recycle?

Design: In this section of the Inquiry Journal form, come up with a plan to keep track of all the garbage you make in one week. What categories of recyclables will you measure? Make sure that you measure each category separately so that you can compare them to each other. How much of your garbage goes to the landfill. How will you estimate the volume of something as lumpy and bumpy as garbage? How much of the garbage that comes out of your home is yours? How will you take one week's worth of information and use it to estimate a year's worth of garbage?

Data Collection: Make a table to keep track of the measurements of your garbage.

Date Presentation: When you have completed gathering data, present your data in graph form (bar graph or pie chart) to show a comparison of how much of each kind of garbage you threw away.

Final Analysis/Conclusion: Your final analysis should discuss the amount of garbage you make in a year. Does more of your garbage go to the landfill or get recycled? Illustrate the number of cubic centimeters of garbage by finding the volume of your classroom. Would your garbage fill the classroom? How many people live in your town? How much garbage would your town make in a year? How many classrooms would be filled? Some types of garbage actually can be recycled. Find out how aluminum, steel, and glass are melted down and used again in a new product.

Looking Backward, Thinking Forward

"Class, quiet please. Mr. Harry Miner is visiting our school to tell us about looking backward and thinking forward. Listen carefully. Mr. Miner is one of the few people left to tell us our history. He was born on January 1, 2000."

"I've seen a lot of change in my 150 years," Mr. Miner started in an excited voice. "But I've learned one thing. It's worth looking backward now to teach you how to think forward. I've done both in my life."

"Tell us about the days when we made mountains of garbage," Melinda asked.

"Well now, let me start at the beginning, the night I was born."

****

(Flashback occurs on the other side of the stage.)

Mother: Look at him, he's perfect. Look, ten toes and ten fingers. He'll be strong and smart and brave. What should we name him?

Father: How about Harry? Harry sounds like the name of a child who will do great things for our world. I know that our little boy, born at the stroke of the new millennium, will grow up to always do good.

****

"So that's how I got my name, Harry Miner."

"But tell us about the garbage! Did they really throw away precious glass, aluminum, and steel back then, Mr. Miner?" asked Lauren, squirming with excitement.

"I'm getting to that," Mr. Miner continued. "You see, my parents taught me to do what was right. So when I was growing up I felt that it was just wrong to throw so much away. Do you know what I decided to do?"

"FIGHT GARBAGE!" the class shouted.

"You're right. It's hard to imagine, but back then people threw away things they'd only used for two minutes."

"You're kidding us," Robert laughed. "I don't believe you."

"I kid you not. We thought backwards. We thought the Earth was there for us to use and that we could throw away anything we wanted."

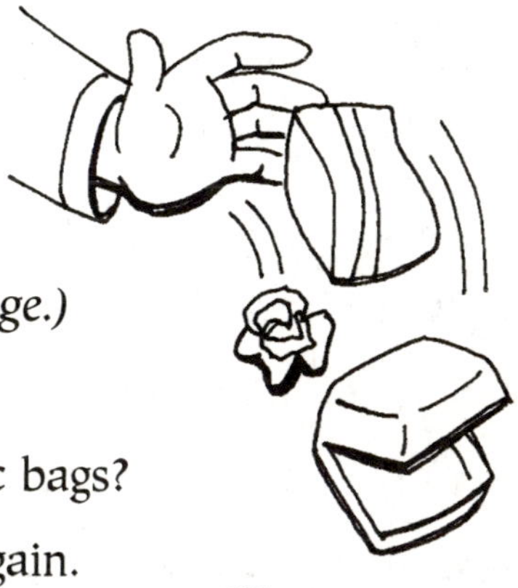

****

(Flashback occurs on the other side of the stage.)

Mother: Help me carry in the groceries, will you Harry?

Young boy, Harry: Sure, Mom. Where should I put these plastic bags?

Mother: Just toss them in the garbage. Nobody can use them again.

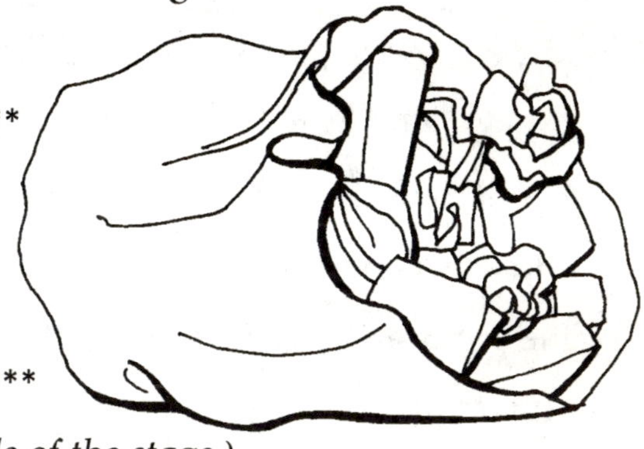

****

The students exploded with laughter.

"And that's not all. Listen."

****

(Flashback occurs on the other side of the stage.)

Mother: Harry, finish up your Wonder Burger Meal. Don't forget to stuff the napkins, French fry bag, the straw, the soda cup, and the from the hamburger into the bag. Then we can throw it all away.

Young boy, Harry: Yes, Mother.

****

"Now tell us about glass, Mr. Miner," Grant begged. "I hear they threw it away, too."

"Just as sure as there's a nose on my face," Mr. Miner continued. "Nobody thought it was worth anything. Our backward thinking meant we thought it was all just junk. Even plastic bottles were trash. Remember, that's back in the days when they made plastic from petroleum. Of course, we used up all the petroleum a hundred years ago. But, back when I was a kid, we didn't ever think about the fact that plastic was made from a rare resource, prehistoric swamps!"

"How did you change their minds, Mr. Miner?" Amelia asked.

"Well, the way you change another person's mind is to change your own mind first. The story of how I changed my mind is embarrassing, but I'll tell it because it's true. When I was a young man, those landfills were huge. The one near my home was 150 feet high. I was down on my luck and didn't have enough money to go out to a movie. So a couple of friends and I went out to explore the landfill. We even found rats there."

"Yuck! Why were rats at the landfill, Mr. Miner?" asked Lauren, crinkling up her nose.

"Good question. The truth is that people threw out food and the rats came to eat."

"But what about the city's compost?" asked Robert. "Didn't people care about making compost for the parks and farms?"

"It was all just backward thinking," said Mr. Miner. "People threw away all of their scraps. The truth is, they were growing rats. That landfill was really just a rat ranch but they didn't think of it that way. Anyway, when I was young man, I changed my mind and this is how."

****

(Flashback occurs on the other side of the stage.)

Young man, Harry: Listen we didn't have money for a movie, so exploring the dump seemed like a good idea.

Friend 1: Well, it wasn't a bad idea until we realized this dump had lots of rats! Let's go.

Friend 2: Yeah, I agree. I had a good time exploring what was thrown away, and I even found a practically new toaster that might work. But now I just want to go home.

Young man, Harry: You're right. Let's go. Wait...do you see that? It looks like a diamond.

Friend 1: Where? Harry, who would throw away a diamond?

Friend 2: It's just a glass bottle. I'm going home before we run into any more rats.

Young man, Harry: You're right, it's just a bottle. Look at it gleaming in the moonlight, though. It sure is a pretty sight.

(Friends leave, Harry stays and looks at the bottle.)

****

"My friends thought I was crazy, but that night I got to thinking about glass. It sure was a pretty bottle in such a wasteful place. I started reading about landfills. The more I learned, the more I realized that we needed to stop discarding valuable materials and start thinking about the future. That's when my forward thinking started. I made a deal with the landfill operator. He let me go out on weekends and mine for glass."

"Were you really the first one to dig into those mountains of glass, plastic, aluminum, and steel, Mr. Miner?" asked Melinda.

"Yes, those were the good old days. All by myself out there, I'd find rich veins of glass, deposits of plastic soda bottles, and islands of cans. I started making a little money at it.

Actually, it was a lot of money. It didn't last, of course."

"When did people realize they should recycle and use it all again?" Lauren asked.

"Well, at first they were just looking for the money."

****

(Flashback occurs on the other side of the stage.)

Mr. Miner: You get your hands off my landfill!

Garbage Collector 1: It doesn't belong to you. Why, I recognize that plastic bottle right over there. (Laughing) Used it last Friday. And this one here is the one I used to fill my baby's bottle week ago Tuesday.

Garbage Collector 2: (Teasing) That's right, I remember it. And those soda cans over there were the ones I used at Junior's fifth birthday party. They're mine.

Mr. Miner: You don't fool me for a minute, all you're thinking about is getting rich fast. I'm in this business to save the world and to get rid of garbage dumps.

****

"What did you do, Mr. Miner?" asked Robert.

"I bought that landfill. Why, I'd saved enough in the few years of mining it to buy it outright. Then I really started using that landfill to its best advantage. Why I ran an electricity-producing plant from burning up the parts of the landfill I didn't want. It didn't produce any air pollution either, because I filtered the smoke."

"What happened then?" asked Amelia.

"Well, I forgot to think ahead. Things were going so well that I thought it would always last. Pretty soon though, the forward thinkers left me behind. You know the story. Scientists invented plants and bacteria that could manufacture biodegradable plastics. Glass, steel, and aluminum became so valuable that recycling centers got all the good stuff before it got thrown away. Cities started composting projects and pretty soon the landfills were gone. There was nothing left for me to mine."

"Mr. Miner," the teacher said. "What is the one piece of advice you could give these children?"

"My life was spent getting rid of the landfill. I won. During my life I've seen people turn the corner from wasters to recyclers. For life to succeed, we can't produce waste faster than we use it. Mother Nature doesn't like that."

"Mr. Miner, do you think that we'll ever go back to the old ways?"

"No. These children know that you need to look backward and think forward."